Applied Condition Monitoring

Volume 1

Series editors

Mohamed Haddar, National School of Engineers of Sfax, Tunisia

Walter Bartelmus, Wroclaw University of Technology, Poland

Fakher Chaari, National School of Engineers of Sfax, Tunisia
e-mail: fakher.chaari@gmail.com

Radoslaw Zimroz, Wroclaw University of Technology, Poland

About this Series

The book series Applied Condition Monitoring publishes the latest research and developments in the field of condition monitoring, with a special focus on industrial applications. It covers both theoretical and experimental approaches, as well as a range of monitoring conditioning techniques and new trends and challenges in the field. Topics of interest include, but are not limited to: vibration measurement and analysis; infrared thermography; oil analysis and tribology; acoustic emissions and ultrasonics; and motor current analysis. Books published in the series deal with root cause analysis, failure and degradation scenarios, proactive and predictive techniques, and many other aspects related to condition monitoring. Applications concern different industrial sectors: automotive engineering, power engineering, civil engineering, geoengineering, bioengineering, etc. The series publishes monographs, edited books, and selected conference proceedings, as well as textbooks for advanced students.

More information about this series at http://www.springer.com/series/13418

Vasile Radu

Stochastic Modeling of Thermal Fatigue Crack Growth

 Springer

Vasile Radu
Structural Integrity Assessment Subsidiary
    of RATEN
Institute for Nuclear Research Pitesti
Mioveni
Romania

ISSN 2363-698X   ISSN 2363-6998   (electronic)
Applied Condition Monitoring
ISBN 978-3-319-38519-8   ISBN 978-3-319-12877-1   (eBook)
DOI 10.1007/978-3-319-12877-1

Springer Cham Heidelberg New York Dordrecht London

Printed on acid-free paper

Springer International Publishing AG Switzerland is part of Springer Science+Business Media
(www.springer.com)

To my parents with gratitude and love

Preface

The assessment of thermal fatigue crack growth due to turbulent mixing of hot and cold coolants presents significant challenges, in particular, to determine the thermal loading spectrum. Thermal striping is defined as a random temperature fluctuation produced by incomplete mixing of fluid streams at different temperatures, and it is essentially a random phenomenon in a temporal sense.

The book's objective is to develop a systematic approach of stochastic modeling to assess thermal fatigue crack growth in mixing tees, based on the power spectral density (PSD) of temperature fluctuation at the inner pipe surface. Based on the analytical solution for temperature distribution through the wall thickness, obtained by the Hankel transform, a frequency, temperature response function is proposed, in the framework of single-input, single-output (SISO) methodology from random noise/signal theory under sinusoidal input. For elastic thermal stresses, developed in a previous work, the magnitude, frequency response function is first derived for hoop stress, and checked against prediction by FEA. The frequency response of the stress intensity factor (SIF) is obtained by the polynomial fitting procedure of stress profiles through the thickness at various instants of time. The variability in load is given by the statistical properties of the thermal spectrum. The temperature spectrum is assumed to be given as a stationary normalized Gaussian narrow-band stochastic process, with constant PSD for a defined range of frequencies. The connection between SIF's PSD and temperature's PSD is assured with SIF frequency response function modulus. The frequency of peaks of any magnitude for KI, which is supposed to be a stationary narrow-band Gaussian process, is characterized by Rayleigh's distribution, and, consequently, the expected value of crack growth rate in respect to cycles is obtained.

The probabilities of failure are estimated by means of the Monte Carlo methods by considering a limit state function, which is based on the proposed stochastic model. The results of the stochastic approach to modeling of thermal fatigue crack growth in mixing tees are completed with probabilistic input to account variability in material characteristics, and finally an application is given to obtain the probability of mixing tees piping failure as a function of the time reference period.

The book addresses to undergraduate students, young scientists and people involved in the practical application of the stochastic and probabilistic fracture mechanics for structural integrity assessment on industrial components.

October 2014   Vasile Radu

Contents

Nomenclature

K_I	stress intensity factor for Mode I
C, n	Paris's law parameters
a	crack size
N	number of stress cycles
$\triangle K$	range of the stress intensity factor
da/dN	increment of fatigue crack advance per stress cycle
$X(t)$	stochastic process
K_{max}	maximum stress intensity factor
R	stress ratio
S	maximum stress level in the loading spectrum
$fX(x)$	probability density function
$E[X]$	expectation of X
μX	mean
$\sigma^2 X$	variance
σX	root mean square
$R_X(\tau)$	autocorrelation function
ω	frequency (radians per second)
$S_X(\omega)$	spectral density
$H(\omega)$	frequency response function
$H^*(\omega)$	complex conjugate of H(ω)
$|H(\omega)|$	magnitude of the frequency response function
λ	thermal conductivity
ρ	mass density
c	specific heat conduction

k	thermal diffusivity
$\Theta(r,t)$	temperature changes from the reference temperature at any radial position r and at time t
$J_0(z)$, $Y_0(z)$	Bessel's functions of first and second kind of order 0
σ_r, σ_θ	the radial and hoop stresses respectively
E	Young modulus
v	Poisson's ratio
α	coefficient of the linear thermal expansion
τ	Gamma function
P_f	probability of failure

Abbreviations

SISO	single-input, single-output methodology
PSD	power spectral density
SIF	stress intensity factor
FEA	finite element analysis
ISI	in-service inspection
LEFM	linear elastic fracture mechanics
PFM	probabilistic fracture mechanics
FFT	Fast Fourier Transform
IFT	Inverse Fourier Transform
RMS	root mean square
DSA	deterministic spectral amplitudes
RSA	random spectral amplitudes
FRF	frequency response function
MCS	Monte Carlo simulation
PDF	probability density function
CDF	cumulative density function
RHRS	Residual Heat Removal System
NPP	Nuclear Power Plant

Chapter 1
Introduction

The assessment of the thermal fatigue damage (crack initiation) and subsequent crack growth due to thermal stresses from turbulent mixing or vortices in industrial piping systems remains a demanding task, and much effort continues to be devoted to experimental, FEA and analytical studies.

The problem of thermal fatigue in mixing areas arises in pipes where a turbulent mixing or vortices produces rapid fluid temperature fluctuations with random frequencies. Structures exposed to such temperature fluctuations can suffer thermal fatigue damage and, subsequently, cracking phenomena, which can produce through wall cracks. Thermal striping is defined as a random temperature fluctuation produced by incomplete mixing of fluid streams at different temperatures. It can arise in a certain light water reactor, but also in certain fast breeder reactor structures, notably those situated above the core, because of the large temperature differences that exist between sodium emerging from the core sub-assemblies and from the breeder sub-assemblies. Other areas of potential occurrence include pressurized water reactor nozzles where stratified flows are encountered. In dry-out zones in steam generators, the fluid/steam boundary can oscillate and induce temperature fluctuations on the component surface [1].

The results in temperature fluctuations can be local or global and induce random variations of local temperature gradients in structural walls of pipe, which lead to be cyclic thermal stresses. These cyclic thermal stresses are caused by oscillations of fluid temperature, and the strain variations result in fatigue damage, cracking and crack growth. In particular, accurate representation of the load is a complex issue, and much effort continues to be devoted to experimental and analytical studies in this area.

Transient temperature response inside of an infinite slab to a sinusoidal surface-temperature input has been investigated by several researchers [2, 3, 4, 5]. For cylindrical geometry, it was used mainly isothermal internal boundary condition and with various types of thermal loading at the outer surface [1, 6, 7]. The determination of the influence of such a random process on subsurface temperatures is important in establishing the proper depth at which temperature sensitive becomes a concern. Utilizing the method of random process theory, it is possible to determine statistical averages such as the mean and standard deviation of the response from the corresponding statistical description of the input process

© Springer International Publishing Switzerland 2015

V. Radu, *Stochastic Modeling of Thermal Fatigue Crack Growth*,

Applied Condition Monitoring 1, DOI: 10.1007/978-3-319-12877-1_1

provided that the governing differential equations are linear. If in addition, the applied random process is normally distributed the output process will also be normal. The effect of spatial incoherence in surface temperature fluctuation can be used to calculate the mean square stresses and the mean square equivalent strain range, that may be used as a measure of crack initiation likelihood [8]. Also, this type of incoherence has the effect of the stress intensity factor in thermal striping. By assuming a perfect spatially coherent, but a temporal incoherence it was developing a method of calculating the crack propagation using linear elastic fracture mechanics and stochastic properties of the temperature spectrum [6].

Thermal striping remains a timely subject in the structural integrity area, also for future fast spectrum reactors [9], mainly with the objective of establishing thermal striping limits or appropriate screening criteria.

The present book proposes an analytical approach of stochastic modeling to assess thermal fatigue crack growth in mixing tees, based on the power spectral density (PSD) of temperature fluctuation at the inner pipe surface. Chapter 2 is devoted to the statistical nature of crack growth. In Chapter 3, a short review of the methods used to apply the stochastic methodology for fatigue crack growth in various applications is done, most of them based on the solving of Fokker-Plank stochastic equation with Markov modeling of processes. The basic mathematical tools used for stochastic fatigue analysis are presented, for easy handling of the stochastic particularities connected with structural assessment of fatigue crack growth under a thermal spectrum. The model of stochastic fatigue crack growth due to temperature fluctuation is developed in Chapter 4, assumed to be a normalized stationary Gaussian random process. Based on the analytical solution for temperature distribution through the wall thickness, obtained by means of the Hankel transform, a frequency, temperature response function is proposed, in the framework of single-input, single-output (SISO) methodology from random noise/signal theory under sinusoidal input. For elastic thermal stresses, developed in a previous work, the magnitude, frequency response function is first derived for hoop stress, and checked against prediction by FEA. The frequency response of the stress intensity factors (SIF), which is considered for a shallow long axial crack on the inner pipe surface, is obtained by the polynomial fitting procedure of stress profiles at various instants of time. The variability in load is given by the statistical properties of a thermal spectrum. The temperature spectrum is assumed to be given as a stationary Gaussian narrow-band stochastic process, with constant PSD for a defined range of frequencies. The connection between SIF- PSD and temperature-PSD are assured with SIF frequency response function modulus. The frequency of peaks of any magnitude for K_I, which is supposed to be a stationary narrow-band Gaussian process, is characterized by Rayleigh's distribution, and, the expected value of crack growth rate in respect to cycles is obtained. To turn on to the crack growth rate in respect to the time variable, it is necessary to introduce the peak crossing rate, that for a harmonic Gaussian process is similar to zero up-crossing rate. The last one is inferred with the aid of the PSD of K_I.

The probabilities of failure are estimated by means of the Monte Carlo methods. Firstly, has defined a limit state function, based on the stochastic model developed. The Civaux case [10] has been chosen to apply the stochastic model from the book, in Chapter 5. The variability in statistical properties of material parameters is usually accounted by the statistical properties of Paris law parameters C and n. Also, the initial crack depth of flaws has a certain probability density function, which is more related to probability of detection based on experimental in-service inspection (ISI) results. A lognormal probability density function of C scaling parameter and an exponential one for initial crack depth are used to provide a probabilistic input for solving the integral giving the crack depth as a function of time.

The results of the stochastic approach to modeling of thermal fatigue crack growth in mixing tee, completed with probabilistic input to account variability in material characteristics, are given as the probability of failure as a function of the time reference period. The Civaux 1 damage case is characterized by a short time for its development to a significant depth through the wall, which was only about ≈ 1500 hours [10]. Metallurgical examinations revealed substantial cracks and also some networks of small thermal fatigue cracks close to the welds, but no fabrication defects. The probability of failure, obtained by analysis with the proposed stochastic model is around 80%, which is a very high value in connection with the nuclear safety management, and in agreement with value above mentioned.

The present methodology based on the stochastic modeling of thermal fatigue crack growth can be used to analyze and improve the screening criteria proposed to avoid cracking issues in industrial piping, especially in tee connection where turbulent mixing of flows with different temperature can occur.

References

[1] Jones, I.S.: The frequency response model of thermal striping for cylindrical geometries. Fatigue and Fracture of Engineering Materials and Structures 20(6), 871–882 (1997)

[2] Heller, R.A.: Temperature response of an infinitely thick slab to random surface temperature. Mech. Res. Comm. 3, 379–385

[3] Kamaya, M.: Crack growth under thermal fatigue loading (effect of stress gradient and relaxation), PVP2009-77547. In: Proceedings of the ASME 2009 Pressure Vessels and Piping Division Conference, July 26-30, Czech Republic, Prague (2009)

[4] Jones, I.S., Lewis, M.W.: A frequency response method for calculating stress intensity factors due to thermal striping loads. Fatigue and Fracture of Engineering Materials and Structures 17(6), 709–720 (1994)

[5] Galvin, B.J., Graham, I.D., Jones, I.S., Rothwel, G.: A comparison between the finite element and frequency response method in the assessment of thermal striping damage. International Pressure Vessels and Piping 74, 205–212 (1997)

[6] Miller, A.G.: Crack propagation due to random thermal fluctuation: effect of temporal incoherence. International Journal of Pressure vessels and Piping 8, 15–24 (1980)

[7] Heller, R.A., Thangjitham, S.: Probabilistic Methods in Thermal Stress Analysis. In: Hetnarski, R.B. (ed.) Thermal Stress II, Elsevier Science Publishers B.V (1987)

[8] Miller, A.G.: Equivalent strain range due to random thermal fluctuations: effect of spatial incoherence. International Journal of Pressure vessels and Piping 8, 105–130 (1980)

[9] Chellapandi, P., Chetal, S.C., Raj, B.: Thermal striping limits for components of sodium cooled fast spectrum reactors. Nuclear Engineering and Design 239, 2754–2765 (2009)

[10] Chapuliot, S., Gourdin, C., Payen, T., Magnaud, J.P., Monavon, A.: Hydro-thermal-mechanical analysis of thermal fatigue in a mixing tee. Nuclear Engineering and Design 235, 575–596 (2005)

Chapter 2
Background on Stochastic Models for Cumulative Damage Process

Abstract. The chapter provides a short review of the stochastic models applied for cumulative fatigue damage in structural components. The empirical fatigue crack growth is given by Paris's law, and it relates the increment of fatigue crack advance, *da/dN*, per stress cycle to range of the stress intensity factor *ΔK* in the framework of linear elastic fracture mechanics (LEFM). Investigation of the randomness of fatigue crack growth rate under service, loading conditions should consider the statistical characteristics of the crack growth law under constant amplitude loadings, and also the randomness of loadings that gives rise to fatigue under variable amplitude loads. Several probabilistic models for crack growth have suggested to "randomize" the deterministic crack propagation by a stochastic process. Few stochastic models of the cumulative fatigue damage process are nominated in this chapter. A method of calculating crack propagation by linear elastic fracture mechanics (LEFM) in the case of Gaussian temperature fluctuations has been proposed by Miller, and the principal steps are outlined. The main features of various structural assessment techniques for thermal striping and thermal fatigue crack growth but also some matters in respect of these methods are shortly reviewed.

2.1 Statistical Nature of the Crack Growth Process

The parameters that affect structural fatigue performance of metallic components may include the applied stress (or loading, in general), the component geometry and structural characteristics, the material properties and operating environment agents. The empirical fatigue crack growth is given by Paris's law in the following form

$$\frac{da}{dN} = C \cdot \left(\Delta K \right)^{n}.$$

$$(2.1)$$

It relates the increment of fatigue crack advance, *da/dN,* per stress cycle to range of the stress intensity factor *ΔK* in the framework of linear elastic fracture mechanics (LEFM). Here *ΔK* is resulted from far-field nominal stress range and

© Springer International Publishing Switzerland 2015

V. Radu, *Stochastic Modeling of Thermal Fatigue Crack Growth,*

Applied Condition Monitoring 1, DOI: 10.1007/978-3-319-12877-1_2

component geometry. The constant C and n are empirical, which depend on the material property and the environment; n is also called the fatigue exponent. The experimental data on fatigue crack growth are recorded as the crack size, a, versus the number of stress cycles, N. Subsequently, the fatigue experimental data are numerical processed as a crack growth rate da/dN versus stress intensity factor range. Due to various sources of uncertainty, the experimental data are usually dispersed is on the regression line. Actually, the regression line only describes the crack growth in the median sense. Investigation of the randomness of fatigue crack growth rate under service, loading conditions should consider the statistical characteristics of the crack growth law under constant amplitude loadings, and also the randomness of loadings that gives rise to fatigue under variable amplitude loads.

Several probabilistic models for crack growth are reviewed in [1], most of them suggested to "randomize" the deterministic crack propagation by a stochastic process $X(t)$:

$$\frac{da(t)}{dt} = X(t) F(a, \Delta K, K_{max}, R, S),\tag{2.2}$$

Where: a is crack depth, ΔK is the stress intensity factor range, K_{max} the maximum stress intensity factor, R, is the stress ratio, S, is the maximum stress level in the loading spectrum, and $a(t)$ is the crack size at time t. The suggestion is frequently to model the process $X(t)$ either by a lognormal stationary stochastic process with a median value equal to unity or by a random pulse train. The reference [2] distinguishes few stochastic models of the cumulative fatigue damage process which are nominated in the next.

a) Random Variable model. This model regards parameters in the crack growth law simply as random variables [3, 4, 5, 6, 7, 8].
b) Markov Chain Model. Bogdanoff and Kozin first proposed this model [9] in which both state and time are handled discretely with a lot of practical applications.
c) Stochastic Differential Equation Model. This is a such model as to treat the crack growth law as a stochastic differential equation by introducing temporal fluctuations into parameters in the law [10, 11, 12, 13, 14, 15, 16].
d) Locally Averaged Model. The model is to discuss statistical properties of whole life by summing up locally averaged lives in order to realize the constraint condition of the constant correlation distance [17].
e) Renewal Process Model. With the application of the stochastic point process, this model represents the state transition in terms of non-homogeneous Poisson processes [9].

All models involve a number of hypotheses (the distribution of time to failure, the nature of the correlation structure of noise, etc.), which limit their applicability. Additionally, developments are required to enhance these probabilistic models and to make them more generally applicable for modeling structural fatigue stochastic.

2.2 Loading Uncertainty

The knowledge of the stress's history, stress amplitude or stress range is required for fatigue analyses. In the first instance, the development of fatigue damage depends on the stress ranges. In many applications, the amplitude of stress range is usually not constant with respect to the time. Modeling the stress ranges for purposes of fatigue reliability analysis involves two parts: definition of cycles based on stress histories and identification of probabilistic characteristics of those cycles.

In the case of a narrow-band process, the stress ranges history is easier to be identified with the individual stress cycles. This is more difficult to be done in the wide band stress process, mainly due to complexity of the frequency content of the whole process. Some approaches were carried out to surpass this difficulty, such as track filtering and rain-flow counting. In the first one, the stress process is transformed into the equivalent narrow band stress process − from the damage accumulation point of view. Therefore, the subsequent fatigue analysis can be conducted as a narrow-band stress fatigue problem. The rain-flow counting approach identifies the stress range from "closed stress-strain loops." This approach is widely used in cumulative fatigue damage based on the Palmgren - Miner's rule (or similarly). It considers that the portion of fatigue damage in the total fatigue life contributed by a certain stress level is linearly proportional to the fraction of the number of cycles corresponding to that stress level. Sometimes the linearly proportionality is replaced with proportionality to the fraction at particular power. The stochastic rain-flow counting is successfully applied to accurate estimation of fatigue damage under broadband random loading [18,19] induced by bimodal processes [20]. The rain-flow counting algorithm is not straightforward to be applied in the original form to the crack growth analysis. An evaluation of the rain-flow cycle counting method for handling probabilistic characteristics of cycles identified should be conducted. Computationally efficient methods for treating stress in time or cycle domains require further investigations.

2.3 Thermal Fatigue Crack Growth Modeling

A method of calculating crack propagation by linear elastic fracture mechanics (LEFM) in the case of Gaussian temperature fluctuations has been, firstly, proposed by Miller [6]. The successive steps in the calculation are:

- For a harmonic surface temperature variation of arbitrary frequency calculates the temperature variation at any depth using the heat conduction equation and the appropriate thermal boundary conditions;

- Calculate the thermal stress at any depth using the linear elastic plate theory and appropriate mechanical boundary conditions;
- Calculate the stress intensity factor (SIF) using LEFM integral method; note that temperature, thermal stress and SIF vary harmonically at the input frequency;
- From the surface, temperature power spectral density (PSD) and SIF response function, calculate the power spectral intensity factor;
- Calculate the frequency distribution of SIF maxima;
- Using a material crack propagation law, calculate the crack propagation rate;
- Integrate the crack propagation rate equation to determine the crack length.

Miller also notes in his approach some assumptions and limitations to be included:

- No account has been taken of the degree of spatial coherence;
- Large scale of plasticity has been ignored (the calculation is restricted to LEFM);
- The statistics are assumed to be Gaussian;
- Linear summation of damage is assumed;
- Crack initiation is not considered.

Clayton and Irvine, [21], described the main features of various structural assessment techniques for thermal striping, but also some matters in respect of these methods are outlined. As analysis methods, they shortly reviewed: method based on the inlet temperature difference, methods based on periodic temperature fluctuations (SIN-method), a method where the temperature spectra is available, the method using Gaussian statistics (Miller approach [6]), and method where a temperature-time traces is present. It is also noted that the thermal stripping produces surface temperature variations which a randomly occurring both in the temporal and spatial senses. Consequently, the size of eddies has a number of influences on the resulting damage.

An approach similar Miller is used also in the paper [5] where the evaluation method for crack propagation due to thermal striping is extended to simulate multiple cracks. The work is supported by FE analysis and thermal fatigue tests.

Thermal fatigue crack growth in a fast breeder reactor is investigated in reference [16]. This is theoretically performed with the aid of probabilistic fracture mechanics (PFM) under conditions:

- The temperature variation is a narrow-band stationary process;
- The crack grows owing only to the peak stress variation.

First, a statistical property of residual life of the component with single crack is derived in an analytical form with the aid of an extended Markov approximation method, which is an efficient mathematical technique in PFM. Next, discussion is carried out on the generalization of the primitive model to the case with plural cracks, where a stress relaxation factor is introduced to express a stress intensity factor of each crack. Finally, a sensitive analysis is performed with respect to some model parameters.

A valuable work on the fatigue crack behavior under thermal stresses is those of Grűter and Huget [22] where is considered the envelopes of stress distribution across the wall. The frequencies of fluid streams are considered in the thermal stress derived for wall thickness in the conservative manner, but not in stochastic fashion.

References

[1] Yao, J.T.P., Kozin, F., et al.: Stochastic fatigue, fracture and damage analysis. Structural Safety 3, 231–267 (1986)

[2] Ishikawa, H., et al.: Reliability assessment of structures based upon probabilistic fracture mechanics. Probabilistic Engineering Mechanics 8, 43–56 (1993)

[3] Kozin, F., Bogdanoff, J.L.: A critical analysis of some probabilistic models of fatigue crack growth. Engineering Fracture Mechanics 18, 59–89 (1981)

[4] Yang, J.N., Manning, S.D.: A simple second order approximation for stochastic crack growth analysis. Engineering Fracture Mechanics 53, 677–686 (1996)

[5] Yang, J.N., Manning, S.D.: Stochastic crack growth analysis methodologies for metallic structures. Engineering Fracture Mechanics 37, 1105–1124 (1990)

[6] Lin, Y.K., Yang, J.N.: On statistical moments of fatigue crack propagation. Engineering Fracture Mechanics 18, 243–256 (1983)

[7] Yang, J.-N.: Simulation of random envelope processes. Journal of Sound and Vibration 21(1), 73–85 (1972)

[8] Wen-Fang, W.: Computer simulation and reliability analysis of fatigue crack propagation under random loading. Engineering Fracture Mechanics 45, 6697–6712 (1993)

[9] Kozin, F., Bogdanoff, J.L.: On the probabilistic modeling of fatigue crack growth. Engineering Fracture Mechanics 18, 623–632 (1983)

[10] Sobczyk, K.: Modeling of random fatigue crack growth. Engineering Fracture Mechanics 24, 609–623 (1986)

[11] Tsurui, A., Tanaka, H., Tanaka, T.: An analytical method for leak before break assessment based upon stochastic fracture mechanics. Nuclear Engineering and Design 147, 171–181 (1994)

[12] Tsurui, A., Tanaka, H., Tanaka, T.: Probabilistic analysis of fatigue crack propagation in finite size specimens. Probabilistic Engineering mechanics 4, 120–127 (1989)

[13] Enneking, T.J., Spencer Jr., B.F., et al.: Computational Aspects of the Stochastic Fatigue Crack Growth Problem. In: Computational Mechanics of Probabilistic and Reliability Analysis, pp. 418–430.

[14] Lei, Y.: A stochastic approach to fatigue crack growth in elastic structural components under random loading. Acta Mechanica 132, 63–74 (1999)

[15] Zhu, W.Q., Lin, Y.K.: On fatigue crack growth under random loading. Engineering Fracture Mechanics 43, 1–12 (1982)

[16] Tanaka, H., Tsurui, A.: Random propagation of a semielliptical surface crack as a bivariate stochastic process. Engineering Fracture Mechanics 33, 787–800 (1989)

[17] Ditlevsen, O.: Random fatigue crack growth – a first passage problem. Engineering Fracture Mechanics 23, 467–477 (1986)

[18] Tovo, R.: Cycle distribution and fatigue damage under broad-band random loading. International Journal of Fatigue 24, 1137–1147 (2002)

[19] Okajima, S., et al.: Fatigue damage evaluation for thermal stripping phenomena using analytical spectrum method, PVP2005-71682. In: Proceedings of PVP 2005, ASME Pressure Vessels and Piping Division Conference, Denver, Colorado USA, July 17-21 (2005)

[20] Low, Y.M.: A method for accurate estimation of fatigue damage induced by bimodal processes. Probabilistic Engineering Mechanics 25, 75–85 (2010)

[21] Clayton, A.M., Irvine, N.M.: Structural Assessment Techniques for Thermal Striping. Journal of Pressure Vessel Technology 109, 283–296 (1987)

[22] Grute, L., Huget, W.: Fatigue behavior under thermal stresses. International Pressure Vessels and Piping 10, 335–359 (1982)

Chapter 3
Basic Mathematical Tools for Stochastic Fatigue Analysis

Abstract. In the chapter is given a short description of basic mathematical tools from stochastic process theory, which will be used to develop the stochastic approach of thermal fatigue crack growth in the high cycle domain. Fatigue analysis is often performed in the time domain, in which all input loading and output stress or strain response are time-based signals. In some situations, however, the response stress and input loading are preferable expressed as frequency-based signals, usually in the form of a power spectral density (PSD) plot. In this case, a system function (or a characteristic of the structural system) is required to relate an input PSD of loading to the output PSD of response. The Fast Fourier Transform (FFT) of a time signal can be used to obtain the PSD of the loading, whereas the Inverse Fourier Transform (IFT) can be used to transform the frequency-based signal to the time-based loading. The transform of loading history between the time domain and frequency domain is subject to certain requirements, as per which the signal must be stationary, random, and Gaussian (normal). Thermal striping is a random phenomenon in a temporal sense. In order to have a better understanding of further developments in the present book, some knowledge of stochastic processes is required.

3.1 Stochastic Processes-Ensembles of Time Series

Any system or structural component produces a specific response under certain excitation. If the excitation or response, $X(t)$, is not predictable, the system is in a random state because the value of $X(t)$ cannot be exactly predicted a priority. It can be described only probabilistically.

A stochastic process, let say $X(t)$ is a time-dependent family of random variables whose possible values at any time are governed by probabilistic laws [1]. It becomes a random variable when index parameter, t, is fixed. The probability density function (PDF) of a time history $X(t)$ can be obtained by calculating its statistical properties. With a given probability density function, $f_X(x),$, some statistical properties of the random process $X(t)$ can be obtained [2]. The mean, μ_X, and the variance, σ^2_X, of the process can be calculated as

© Springer International Publishing Switzerland 2015

V. Radu, *Stochastic Modeling of Thermal Fatigue Crack Growth,*

Applied Condition Monitoring 1, DOI: 10.1007/978-3-319-12877-1_3

$$\mu_X = \int_{-\infty}^{+\infty} x \cdot f_X(x)dx \cong \frac{1}{T}\int_0^T X(t)dt , \tag{3.1}$$

$$\sigma_X^2 = \int_{-\infty}^{+\infty} (x-\mu_X)^2 f_X(x)dx \cong \frac{1}{T}\int_0^T [X(t)-\mu_X]^2 dt , \tag{3.2}$$

where T is the record length. When $\mu_X=0$, then σ_X is the root mean square (RMS) of the random process $X(t)$. The RMS is a measure of the amplitude of the process. The process $X(t)$ is called Gaussian if its PDF $f_X(x)$ follows the bell-shape distribution, and PDF is given by

$$f_X(x) = \frac{1}{\sqrt{2\pi}\sigma_X} \exp\left[-\frac{1}{2}\left(\frac{x-\mu_X}{\sigma_X}\right)^2\right], \tag{3.3}$$

where μ_X and σ_X are the mean and standard deviation of the process, respectively.

A random process (or sometime called the stochastic process) may be thought as a "sample" function of time such as $X_1(t)$ in Figure 3.1. [3] which may represent a fluctuation of thermal stresses in mixing tees. As it can be seen, the second component, $X_2(t)$, may undergo with similar, but not identical, stress history. Thus, a collection of an infinite number of sample time histories, such as $X_1(t)$, $X_2(t)$, $X_3(t)$, etc., (Figure 3.1), makes up the random process $X(t)$. In engineering, the ensemble of a sufficiently large number of sample time histories approximates the unlimited collection representing a random process.

Instead of being measured along a single sample, the ensemble statistical properties are determined across the ensemble as shown in Figure 3.1. In this case the amplitudes, $X_i(t)$, of such sample functions are examined, all at the time, t_1, and a density function, $f_{X(t1)}(X(t_1))$, may describe the statistics of the ensemble at t_1. The mean, variance and other moments of $X(t_1)$ are then calculated. The density function may also be evaluated at t_2 as can the moments of $X(t_2)$. The joint density of amplitudes at two-time values t_1 and t_2 separated by a time interval $\tau=t_2-t_1$ is denoted as

$$f_{X(t_1)X(t_2)}\left(X(t_1), X(t_2)\right). \tag{3.4}$$

For a Gaussian process, the ensemble probability density at each time instant and at any two-time, units must be Gaussian. If the expected value of the product

$$X(t_1)\cdot X(t_2) = X(t_1)\cdot X(t_1+\tau), \tag{3.5}$$

is a function of delay time τ and not of running time t as

$$E\left[x(t_1)\cdot x(t_1+\tau)\right] = R(\tau), \tag{3.6}$$

then the process is called weakly stationary [1, 2, 3]. For such a process, the first-order density function $f_X(x)$ is independent of time and all- first-order ensemble averages such as the mean and standard deviation, and root mean squares are time independent. The process is strongly stationary when higher-order moments are also independent of running time.

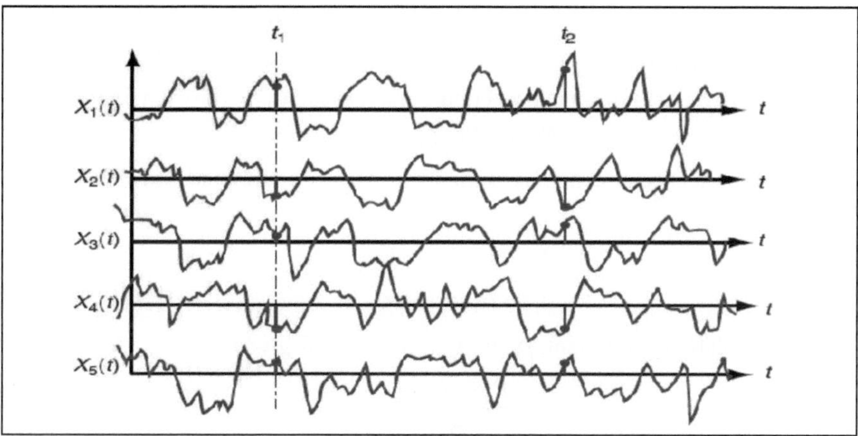

Fig. 3.1 The collection of sample time histories from a random process

Ensemble averaging is carried out across many series at times. This procedure has led to the definition of stationary. A time series can also be averaged along the time axis, resulting in temporal averages. The temporal mean is given by

$$\langle X(t) \rangle = \lim_{T \to \infty} \frac{1}{T} \int_{-T/2}^{T/2} X(t)dt, \tag{3.7}$$

while the temporal mean square is written as

$$\langle X^2(t) \rangle = \lim_{T \to \infty} \frac{1}{T} \int_{-T/2}^{T/2} X^2(t)dt \cdot \tag{3.8}$$

The temporal autocorrelation function is

$$\phi(\tau) = \lim_{T \to \infty} \frac{1}{T} \int_{-T/2}^{T/2} X(t) \cdot X(t+\tau)dt = \langle X(t) \cdot X(t+\tau) \rangle \cdot \tag{3.9}$$

If all temporal averages are equal to the ensemble averages, the process is called "ergotic." Alternatively, in other words, a stationary process is called ergotic if the statistical properties along any single sample are the same as the properties taken across the ensemble. This means that each sample is the same as the properties taken across the ensemble. If a random process is ergotic, it must be stationary; the converse is not true: a stationary process is not necessary ergotic.

3.1.1 The Autocorrelation Function $R_X(\tau)$

The function $R_X(\tau)$, from Equation 3.6, is the "autocorrelation function" of the random process $X(t)$. When $\tau=0$ the autocorrelation function is

$$R_X(\tau=0) = E[X(t_1) \cdot X(t_2)] = \sigma_X^2 + \mu_X^2 = E\left[X^2\right], \qquad (3.10)$$

and $R(0)$ becomes the mean square value of the process when $\mu_X=0$. For $\tau \to \infty$, $R(\infty) \to \mu_x^2$ if $R(\tau)$ does not contain a periodic component. $R(\tau)$ is an even function of τ

$$R(\tau) = R(-\tau). \qquad (3.11)$$

Furthermore, because $X(t)$ is stationary, its mean and standard deviation are independent of t. Thus,

$$E\left[X(t_1)\right] = E\left[X(t_2)\right] = \mu_X, \qquad (3.12)$$

$$\sigma_X(t_1) = \sigma_X(t_2) = \sigma_X. \qquad (3.13)$$

The correlation coefficient, ρ, for $X(t_1)$ and $X(t_2)$ is given by

$$\rho = \frac{R_X(\tau) - \mu_X^2}{\sigma_X^2}. \qquad (3.14)$$

With $\rho=+/- 1$, there is a perfect correlation between $X(t_1)$ and $X(t_2)$, and for $\rho=0$, there is no correlation. Usually the values of ρ are between -1 and +1, the autocorrelation function is

$$\mu_X^2 - \sigma_X^2 \le R(\tau) \le \mu_X^2 + \sigma_X^2. \qquad (3.15)$$

Figure 3.2 shows the property of autocorrelation function $R(\tau)$ of a stationary process $X(t)$. When the time interval (goes to infinity, the random variables at t_1 and t_2 are not correlated.

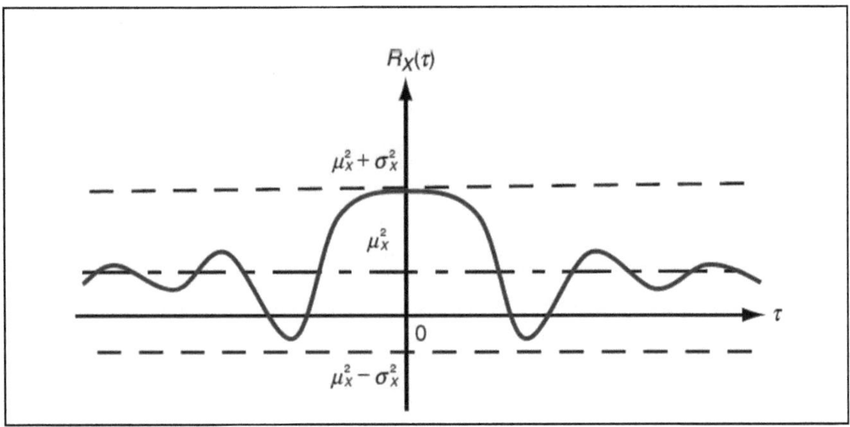

Fig. 3.2 Autocorrelation function of a stationary process

3.1.2 *Fourier Analysis and Power Spectral Density $S_X(\omega)$*

Any periodic time history can be represented by summation of a series of sinusoidal waves of various amplitude, frequency, and phase. If $X(t)$ is a periodic function of time with a time period T, $X(t)$ can be expressed by an infinite trigonometric series of the following form:

$$X(t) = A_0 + \sum_{k=1}^{\infty} \left(A_k \cos \frac{2\pi kt}{T} + B_k \sin \frac{2\pi kt}{T} \right), \tag{3.16}$$

where

$$A_0 = \frac{1}{T} \int_{-T/2}^{T/2} X(t)dt, \tag{3.17}$$

$$A_k = \frac{2}{T} \int_{-T/2}^{T/2} X(t)\cos \frac{2\pi kt}{T} dt, \tag{3.18}$$

$$B_k = \frac{2}{T} \int_{-T/2}^{T/2} X(t)\sin \frac{2\pi kt}{T} dt. \tag{3.19}$$

The Fourier series can be also expressed by using complex coefficients as

$$X(t) = \sum_{k=1}^{\infty} C_k e^{\frac{i 2\pi kt}{T}}, \tag{3.20}$$

where the complex coefficients C_k is given by

$$C_k = \frac{1}{T} \int_{-T/2}^{T/2} X(t) e^{-\frac{i 2\pi kt}{T}} dt. \tag{3.21}$$

The Fourier transform can be considered as the limit of the Fourier series of $X(t)$ as T approaches infinity.

The Equation 3.20 is rewritten

$$X(t) = \lim_{T \to \infty} \sum_{k=-\infty}^{\infty} \left(\frac{1}{T} \int_{-T/2}^{T/2} X(t) e^{-\frac{i 2\pi kt}{T}} dt \right) e^{\frac{i 2\pi kt}{T}}. \tag{3.22}$$

The frequency of the kth harmonic, ω, in radians per second, is

$$\omega_k = \frac{2\pi k}{T}, \tag{3.23}$$

and the spacing between two adjacent periodic functions, $\Delta\omega$, is

$$\Delta\omega = \frac{2\pi}{T}. \tag{3.24}$$

Using Equation 3.24 in 3.22 it becomes

$$X(t) = \lim_{T\to\infty} \sum_{k=-\infty}^{\infty} \left(\frac{\Delta\omega}{2\pi} \int_{-T/2}^{T/2} X(t)e^{-ik\Delta\omega t}dt \right) e^{ik\Delta\omega t}. \tag{3.25}$$

If T approaches to infinity, the frequency spacing, $\Delta\omega$, becomes infinitesimally small, denoted by $d\omega$, and the sum becomes an integral. Thus, Equation 3.25 is expressed by the Fourier transform pair $X(t)$ and $X(\omega)$:

$$X(\omega) = \frac{1}{2\pi} \int_{-\infty}^{\infty} X(t)e^{-i\omega t}dt, \tag{3.26}$$

$$X(t) = \int_{-\infty}^{\infty} X(\omega)e^{-i\omega t}d\omega. \tag{3.27}$$

The function $X(\omega)$ is the forward Fourier transform of $X(t)$, and $X(t)$ is the inverse Fourier transform of $X(\omega)$. The Fourier transforms exits if the following conditions are met:

-    The integral of absolute function exists, i.e.:
-

$$\int_{-\infty}^{\infty} |X(t)|dt < \infty. \tag{3.28}$$

-    Any discontinuities are finite.

Usually the Fourier transform of a stationary random process $X(t)$ does not exist because the condition (3.28) is not met. However, the Fourier transform of the autocorrelation function $R_X(\tau)$ always exists. If the stationary random process $X(t)$ is adjusted (or normalized) to a zero-mean value, i.e.,

$$R_X(\tau \to \infty) = 0, \tag{3.29}$$

the condition

$$\int_{-\infty}^{\infty} \left| R_X(\tau) \right| dt < 0 \tag{3.30}$$

is met. In this case, the pair Fourier transforms of $R_X(t)$ are given by

$$S_X(\omega) = \frac{1}{2\pi} \int_{-\infty}^{\infty} R_X(\tau) e^{-i\omega\tau} d\tau, \tag{3.31}$$

$$R_X(\tau) = \int_{-\infty}^{\infty} S_X(\omega) e^{i\omega\tau} d\omega. \tag{3.32}$$

Here the $S_X(\omega)$ is called the spectral density (or power spectral density, PSD) of the normalized stationary random process $X(t)$. With $\tau{=}0$ Equation (3.32) gives the variance in the form

$$E\left[X^2 \right] = R_X(0) = \int_{-\infty}^{\infty} S_X(\omega) d\omega = \sigma_X^2. \tag{3.33}$$

The Equation 3.33 signifies that the square root of the area under a spectral density plot $S_X(\omega)$ is the root mean square (RMS) of a normalized stationary random process. $S_X(\omega)$ is also called mean square spectral density (Figure 3.3). The negative frequencies have no physical meaning. The most common practice is to consider the frequency from zero to infinity, and to have the frequency expressed in hertz (cycle/second) rather than radians/second. Therefore, the two-sided spectral density, $S_X(\omega)$, can be transformed into an equivalent one-sided spectral density, $W_X(f)$ as follows:

$$E\left[X^2 \right] = \sigma_X^2 = \int_0^{\infty} W_X(f) df, \tag{3.34}$$

where

$$W_X(f) = 4\pi S_X(\omega) \tag{3.35}$$

is the power spectral density (PSD), with $f = \omega/2\pi$ in hertz.

The following spectral density relationship exists for first and second derivatives of a stationary random process $X\,(t)$:

$$S_{\dot{X}}(\omega) = \omega^2 S_X(\omega),\tag{3.36}$$

$$W_{\dot{X}}(f) = (2\pi)^2 f^2 W_X(f),\tag{3.37}$$

$$\sigma_{\dot{X}}^2 = \int_{-\infty}^{\infty} S_{\dot{X}}(\omega)d\omega = \int_{-\infty}^{\infty} \omega^2 S_X(\omega)d\omega = (2\pi)^2 \int_0^{\infty} f^2 W_X(f)df,\tag{3.38}$$

$$S_{\ddot{X}}(\omega) = \omega^4 S_X(\omega),\tag{3.39}$$

$$W_{\ddot{X}}(f) = (2\pi)^4 f^4 W_X(f),\tag{3.40}$$

$$\sigma_{\ddot{X}}^2 = \int_{-\infty}^{\infty} S_{\ddot{X}}(\omega)d\omega = \int_{-\infty}^{\infty} \omega^4 S_X(\omega)d\omega = (2\pi)^4 \int_0^{\infty} f^4 W_X(f)df,\tag{3.41}$$

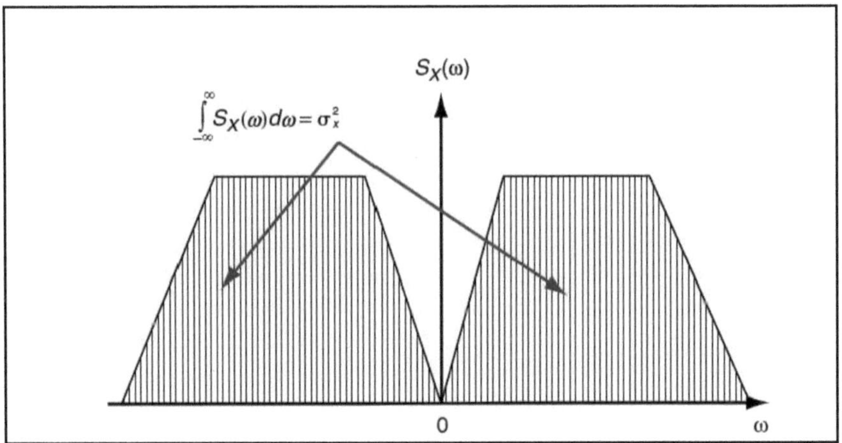

Fig. 3.3 Relationship between the spectral density and root mean square of a normalized stationary random process

Usually, a stationary random process is called a narrow-band process if its PSD has relevant values into a limited frequency interval (narrow band of frequencies), Figure 3.4 (a). The sample realizations of such a process are constituted in practice by cycles of variable range symmetrically placed with respect to the mean, Figure 3.4 (b).

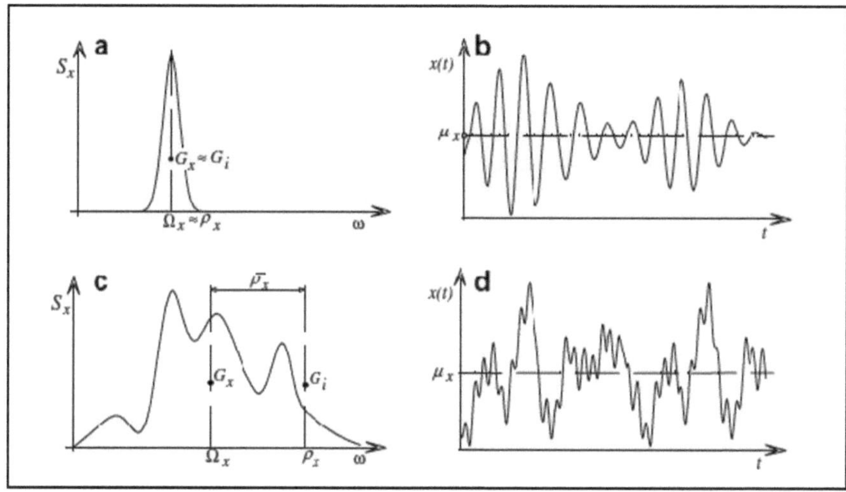

Fig. 3.4 PSD and sample realization of a narrow-band process (a, b) and wide-band process (c, d)

On the contrary, a stationary random process is called wideband, if its PSD has significant values into an interval of frequency (Figure 3.4 (c)). The sample realizations of such a process contain a large number of peaks between two successive mean level crossings (Figure 3.4 (d)).

3.2 Few Special Stochastic Processes

3.2.1 A Stationary Gaussian Process

A stochastic process is called Gaussian if for any integer n and any subset $t_1, t_2,...,$ t_n of T the random variables $X(t_1)$, $X(t_2)$, ..., $X(t_n)$ has a joint Gaussian distribution. The Gaussian process is completely determined by its mean and autocorrelation function, and all the linear transformation of $X(t)$ is also Gaussian. Moreover, the derivative of a Gaussian process is still Gaussian. The original process and its first derivative are independent. The original process and its second derivative have a correlation coefficient equal to the spectral width parameter γ. If $X(t)$ has mean zero, all moments can be expressed in terms of second order moments [4]:

$$E\left(\prod_{s=1}^{n} X_s\right) = 0, \, n=\text{odd}, \tag{3.42}$$

$$E\left(\prod_{s=1}^{n} X_s\right) = \sum E\left[\left(X_{i1}X_{i2}\right)\right]...E\left[\left(X_{in-1}X_{in}\right)\right], \, n=\text{even}, \tag{3.43}$$

The summation is over-all possible ways of combining pairs.

The peak distribution is given by [4]

$$F_p(u) = \Phi\left(\frac{u-\mu_X}{\sigma_X\sqrt{1-\alpha_2^2}}\right) - \alpha_2 \exp\left(-\frac{(u-\mu_X)}{2\sigma_X^2}\right)\Phi\left(\frac{\alpha_2(u-\mu_X)}{\sigma_X\sqrt{1-\alpha_2^2}}\right) \quad (3.44)$$

When study involves time series, it is necessary to generate a Gaussian process, $X(t)$, from a power spectral density. There are two widely used approaches, involving deterministic spectral amplitudes (DSA) and random spectral amplitudes (RSA). Given a two-sided power spectral density $S(\omega)$ of a Gaussian process, $X(t)$, a sample function of discrete sequence, $X(t_i)$, i=0,..., N-1, can be obtained by a DSA method [4]

$$X(t_i) = \sqrt{2}\sum_{k=0}^{N-1} A_k \cos(\omega_k t_i + \phi_k), \quad (3.45)$$

in which

$$A_k = \sqrt{2S(\omega_k)\Delta\omega}, \quad (3.46)$$

$$\omega_k = k\Delta\omega = k\frac{\omega_u}{N}. \quad (3.47)$$

Here ω_u is the cutoff frequency, and ϕ_k is uniform variations within U [0, 2].
A sample function of $X(t_i)$ can be also generated by RSA method [4],

$$X(t_i) = \sqrt{2}\sum_{k=0}^{N-1} A_k \cos(\omega_k t_i + \phi_k), \quad (3.48)$$

with

$$A_k = \sqrt{R_k S(\omega_k)\Delta\omega}, \quad (3.49)$$

$$\omega_k = k\Delta\omega = k\frac{\omega_u}{N}. \quad (3.50)$$

Here ω_u are the cut-off frequency; Rk is of Rayleigh's distribution, and ϕ_k is uniform variations within U[0,2π]. Note that the DSA only consists of one random factor, the random phase, and the amplitudes of A_i deterministic. As a result, the realizations generated from DSA are artificially regular, and each realization has the same variance as the true process. This is not realistic for finite length realizations. Furthermore, $x(t_i)$ is bounded as long as the number of summation terms to generate the sample is limited. In other words, the DSA method may not be able to reproduce the extremes of the process unless the value of N is large enough. The RSA method consists of two random factors, the random phase and the random amplitude. A realization from RSA is more accurate reflection of the irregularity of real processes, especially when the mth power of amplitudes of a process is a major interest. Moreover, the process generated by RSA method is always Gaussian.

3.2.2 *Wiener Process*

A Wiener process $W(t)$ is defined as a process with independent and stationary Gaussian increments,

$$\Delta W = W\left(t + \Delta t\right) - W(t),\tag{3.51}$$

with

$$E\left[\Delta W\right] = 0,\tag{3.52}$$

and

$$E\left[\left(\Delta W\right)^2\right] = \Delta t.\tag{3.53}$$

$W(t)$ is a Gaussian stochastic process with zero mean and autocorrelation function

$$E\left[W\left(t\right)W\left(t + \Delta t\right)\right] = t.\tag{3.54}$$

Almost all sample functions of $W(t)$ are of unbounded variation in every finite interval.

3.2.3 *White Noise*

A white-noise process, $\xi(t)$, is a stationary process with zero mean and autocorrelation function $2\pi S_0 \delta(t)$, in which $\delta(x)$, is a Dirac delta function. Its (two sided) power spectral density is a constant, S_0, over-all frequencies.

3.2.4 *Markov and Diffusion Process*

A Markov process $X(t)$ is one in which, given all past and present states; the knowledge of future states is dependent only on the current state. The evolution of a Markov process is described by a transition probability density function, $f(y,t/x,s)$, which represents the probability density of $X(t)=y$, given that $X(s)=s$, $t>s$. In other words; all probabilistic information of a Markov process can be determined from its transition probability density function and initial state.

A special case of the Markov process is the diffusion process. A Markov process $X(t)$ is called a diffusion process if, for any $\varepsilon > 0$ [1]

$$\lim_{t \to s} \frac{1}{t - s} \int_{|y - x| > \varepsilon} f\left(y, t \mid x, s\right) dy = 0,\tag{3.55}$$

and there exist $m(x,s)$ and $\sigma(x,s)$, such that

$$\lim_{t \to s} \frac{1}{t-s} \int_{|y-x| \le \varepsilon} (y-x) f\left(y,t \,|\, x,s\right) dy = m(x,s), \tag{3.56}$$

$$\lim_{t \to s} \frac{1}{t-s} \int_{|y-x| \le \varepsilon} (y-x)^2 f\left(y,t \,|\, x,s\right) dy = \sigma^2(x,s). \tag{3.57}$$

The condition (3.57) means that a large change in $X(t)$ over a short period of time is impossible. The parameter $m(x,s)$ is called the drift coefficient, and it describes the mean velocity of the increment $X(t)$-$X(s)$ under the condition $X(s)=x$. The diffusion coefficient, $\sigma^2(x,s)$, gives the local magnitude of the fluctuation of $X(t)$-$X(s)$ about the mean value. It can be shown that

$$X(t) - X(s) \simeq m\left(X(s),s\right)\left(t-s\right) + \sigma\left(X(s),s\right)\left(W(t) - W(s)\right), \tag{3.58}$$

where the increments of a Wiener process, $W(t)$-$W(s)$, has the Gaussian distribution, $N(0,(t-s))$.

A diffusion process can be written as the solution of the Ito stochastic differential equation,

$$dX(t) = m\left(X,t)dt\right) + \sigma\left(X,t\right) dW(t). \tag{3.59}$$

Analogously, an n-dimensional vector Markov diffusion process $X(t)$ may be generated from a vector Ito differential equation

$$d\mathbf{X}(t) = \mathbf{M}\left(\mathbf{X},t\right) dt + \mathbf{\Gamma}\left(\mathbf{X},t\right) d\mathbf{W}(t). \tag{3.60}$$

where $\mathbf{M}$ is an n-dimensional drift vector $\mathbf{\Gamma}\mathbf{\Gamma}^{\mathbf{T}}$ is a $n{\times}n$ diffusion matrix, and $\mathbf{W}$ is a vector of n independent Wiener processes. Note that it is not necessary that all components of the vector be Markovian for the vector to be Markovian.

The transition probability density $f(y,t/\,x,s)$, of a Markov diffusion process is uniquely determined by the drift vector and diffusion matrix. Assuming f is continuous with respect to s and the derivatives $\partial f / \partial x_i$ and $\partial^2 f / \partial x_i \partial x_j$ exist and are continuous with respect to s, then f is a solution of the Kolmogorov backward equation is

$$\frac{\partial f}{\partial s} + \sum_{i=1}^{n} m_i(x,s) \frac{\partial f}{\partial x_i} + \frac{1}{2} \sum_{i=1}^{n} \sum_{j=1}^{n} b_{ij}(x,s) \frac{\partial^2 f}{\partial x_i \partial x_j} = 0. \tag{3.61}$$

If $X(t)$ is a homogeneous process, that is, the transition probability is dependent on $\tau=t$-s, not s, then

$$-\frac{\partial f}{\partial \tau} + \sum_{i=1}^{n} m_i \frac{\partial f}{\partial x_i} + \frac{1}{2}\sum_{i=1}^{n}\sum_{j=1}^{n} b_{ij} \frac{\partial^2 f}{\partial x_i \partial x_j} = 0. \tag{3.62}$$

If f is continuous with respect to t and the derivatives, $\partial\left(m_i(x,t)f\right)/\partial x_i$ and $\partial^2\left(b_{ij}(x,t)f\right)/\partial x_i \partial x_j$ exist and are continuous, then f is a solution of the Kolmogorov (or Fokker-Plank) equation

$$\frac{\partial f}{\partial t} + \sum_{i=1}^{n} \frac{\partial\left(m_i(x,t)f\right)}{\partial x_i} - \frac{1}{2}\sum_{i=1}^{n}\sum_{j=1}^{n} \frac{\partial^2\left(b_{ij}(x,t)f\right)}{\partial x_i \partial x_j} = 0. \tag{3.63}$$

3.3 Stationary Gaussian Narrow-Band Process

The mean up-crossing rate and peak distribution are very useful descriptors for structural fatigue loading, and will be used in the stochastic modeling of thermal fatigue crack growth in the following chapter.

The mean up crossing rate $v^+(u,t)$ is the measure of the average frequency that a process crosses a certain level u with positive slope. In other words, it is the mean rate of occurrence of the event $E = \left(X(t)=u \cap \dot{X}(t) > 0\right)$ at time t.

Assuming that the joint density function of $X(t)$ and $\dot{X}(t)$ is $f_{X\dot{X}}(u,v)$, the following expression is the level crossing rate of a stationary random process:

$$v_u^+ = \int_0^{\infty} v f_{X\dot{X}}(u,v)\,dv. \tag{3.64}$$

If $X(t)$ is a Gaussian stationary random process, the expected up crossing rate of $x=u$ is [4]

$$v_u^+ = \frac{1}{2\pi}\frac{\sigma_{\dot{X}}}{\sigma_X}\exp\left(-\frac{u^2}{2\sigma^2}\right). \tag{3.65}$$

The expected rate of zero up-crossing $\dot{N}_{X,0}$ is found by considering $u=0$ in Equation 3.65:

$$\dot{N}_{X,0} = E[v_0^+] = \frac{1}{2\pi}\frac{\sigma_{\dot{X}}}{\sigma_X}. \tag{3.66}$$

By using Equations 3.34 and 3.38, based on one-side PSD into Equation 3.66, the expected rate of zero is

$$\dot{N}_{X,0} = E[v_0^+] = \frac{\sqrt{\int_0^\infty f^2 W_X(f)\,df}}{\sqrt{\int_0^\infty W_X(f)\,df}} \cdot \tag{3.67}$$

The expected rate of peak crossing, $\dot{N}_{X,p}$ is found from a similar analysis of the velocity process $\dot{X}(t)$. The rate of zero down crossing of the velocity process corresponds to the occurrence of a peak in $\dot{X}(t)$, which means the occurrence of the event $E = \left(\dot{X}(t) = 0 \cap \ddot{X}(t) < 0\right)$. The result of a Gaussian random stationary process is

$$\dot{N}_{X,p} = E[v_p] = \frac{1}{2\pi}\frac{\sigma_{\ddot{X}}}{\sigma_{\dot{X}}} \cdot \tag{3.68}$$

By using Equations 3.38 and 3.41 the expected rate of peak crossing becomes

$$\dot{N}_{X,p} = E[v_p] = \frac{\sqrt{\int_0^\infty f^4 W_X(f)\,df}}{\sqrt{\int_0^\infty f^2 W_X(f)\,df}} \cdot \tag{3.69}$$

A Gaussian stationary narrow-band process is smooth and harmonic. For every peak, there is a corresponding zero up-crossing, which means

$$E[v_0^+] - E[v_p], \tag{3.70}$$

or

$$\dot{N}_{X,0} = \dot{N}_{X,p} \cdot \tag{3.71}$$

A wide band process is more irregular, and a measure of this irregularity is the ratio of the zero up-crossing rates to the peak crossing rate. This ratio is called the irregularity factor, γ, in the form

$$\gamma = \frac{E[v_0^+]}{E[v_p]} \cdot \tag{3.72}$$

When $\gamma \rightarrow \infty$ there is an infinite number of peaks for every zero up-crossing and this is considered a wide band random process. The value of $\gamma = 1$ corresponds to one peak per one zero up-crossing, and it represents a narrow-band stationary random process. Alternatively, a narrow- or wide band process can be assessed by the width of its spectrum, which is characterized with the spectral width parameter, λ, defined as

$$\lambda = \sqrt{1 - \gamma^2}$$
(3.73)

Note that $\lambda \rightarrow 0$ represents a narrow-band stationary random process.

The same descriptors for structural fatigue loading can be defined by using the so-called spectral moments M_j of the PSD of a stationary random process (Figure 3.1), which is defined as

$$M_j = \int_0^\infty f^j W_{S_a}(f)\,df \,.$$
(3.74)

In particular, the zero-order moment M_0 provides an area of the PSD that coincides with the variance σ_X^2 of the normalized stationary random process. A family of bandwidth parameters, measures of bandwidth, is described as

$$\alpha_m = \frac{M_m}{\sqrt{M_0 M_{2m}}} \,,$$
(3.75)

which has their values between [0,1]. The one that is most widely used in practice is α_2.

The zero up-crossing rates and the peak crossing rate are given by

$$E[\upsilon_0^+] = \sqrt{\frac{M_2}{M_0}} \,,$$
(3.76)

and

$$E[\upsilon_p] = \sqrt{\frac{M_4}{M_2}} \,.$$
(3.77)

Furthermore, the irregularity factor, γ, and the spectral width parameter, λ become

$$\gamma = \alpha_2 = \sqrt{\frac{M_2^2}{M_0 M_4}} \,,$$
(3.78)

and

$$\lambda = \sqrt{1 - \frac{M_2^2}{M_0 M_4}} \, , \qquad (3.79)$$

respectively.

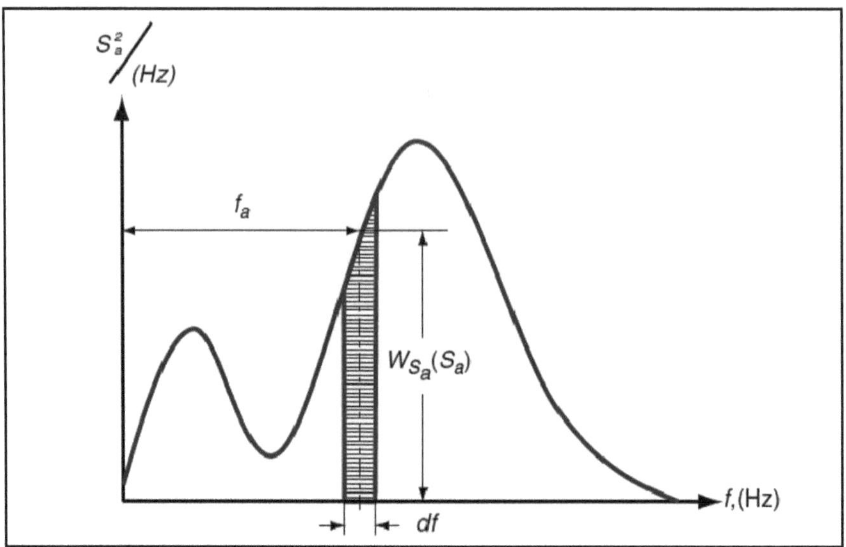

Fig. 3 5 Moments of one-side PSD

The distribution of amplitudes, A, for a stationary narrow-band process Gaussian process, is characterized by the Rayleigh distribution. The probability density function for amplitudes follows a Rayleigh distribution [5] of the form

$$f_A(a) = \frac{a}{\sigma_x^2} \exp\left[-\frac{1}{2}\left(\frac{a}{\sigma_x^2} \right) \right] \qquad (3.80)$$

where a is the random amplitude of the process and σ_x is the mode of Rayleigh's distribution, but also equal to the standard deviation of the underlying Gaussian process. The probability of encountering amplitude greater than a given value a, is the integral of Equation 3.80,

$$P[A > a] = \int_a^\infty \frac{a}{\sigma_x^2} \exp\left[-\frac{1}{2}\left(\frac{a}{\sigma_x} \right)^2 \right] da = \exp\left[-\frac{1}{2}\left(\frac{a}{\sigma_x} \right)^2 \right] . \qquad (3.81)$$

The mth moments of Rayleigh are given by

$$\mu_m = \sigma^m 2^{\frac{m}{2}} \Gamma\left(1+\frac{1}{m}\right),$$

(3.82)

where Γ is the Gamma function. The mean and variance of Rayleigh's distribution are

$$E[a] = \sigma_x \sqrt{\frac{\pi}{2}} \approx 1.253 \sigma_x,$$

(3.83)

and

$$\sigma^2 = \frac{4-\pi}{2} \sigma_x^2 \approx 0.429 \sigma_x^2.$$

(3.84)

The features of Rayleigh's distribution will be used for SIF statistical properties.

3.4 Frequency Response Functions

3.4.1 The General Methodology

Let be $L[Y(t)]$ a linear differential equation as

$$L = \sum_{k=0}^{n} a_k \frac{d^k}{dt^k},$$

(3.85)

with deterministic coefficients, a_k. When the input-output relation is governed by a linear differential equation of form [7]

$$L[Y(t)] = X(t),$$

(3.86)

with $X(t)$ and $Y(t)$ the input and output respectively, the two functions, the frequency may be defined. Letting $H(\omega) \cdot e^{i\omega t}$ be the steady-state response of the system to a unit-amplitude sinusoidal excitation of form

$$X(t) = e^{i\omega t},$$

(3.87)

then $H(\omega)$ is called the frequency response function, and the output becomes

$$Y(\omega) = H(\omega)e^{i\omega t}.$$

(3.88)

By substituting Equation 3.88 into Equation 3.85, the results are

$$\sum_{k=0}^{n} a_k (i\omega)^k H(\omega)e^{i\omega t} = e^{i\omega t} , \tag{3.89}$$

from which

$$H(\omega) = \frac{1}{\displaystyle\sum_{k=0}^{n} a_k (i\omega)^k} . \tag{3.90}$$

Here $H(\omega)$ is a complex function of frequency ω. If $X(t)$ is the real part of the excitation

$$X(t) = \mathrm{Re}[e^{i\omega t}] = \cos \omega t , \tag{3.91}$$

then

$$Y(t) = \mathrm{Re}[H(\omega)e^{i\omega t}] = |H(\omega)|\cos(\omega t + \phi) , \tag{3.92}$$

where

$$\phi = \tan^{-1} \frac{\mathrm{Im}\big[H(\omega)\big]}{\mathrm{Re}\big[H(\omega)\big]} . \tag{3.93}$$

$X(\omega)$ and $Y(\omega)$ are the Fourier transform of $X(t)$ and $Y(t)$:

$$X(\omega) = \int_{-\infty}^{\infty} X(t)e^{-i\omega t}dt , \tag{3.94}$$

$$Y(\omega) = \int_{-\infty}^{\infty} Y(t)e^{-i\omega t}dt . \tag{3.95}$$

If we apply the Fourier transform

$$\sum_{k=0}^{n} a_k \frac{d^k}{dt^k}[Y(t)] = X(t) , \tag{3.96}$$

the result is

$$\sum_{k=0}^{n} a_k Y(\omega)(i\omega)^k = X(\omega) . \tag{3.97}$$

Therefore

$$Y(\omega) = \frac{X(\omega)}{\sum_{k=0}^{n} a_k (i\omega)^k} .$$

(3.98)

By using Equation 3.90 into 3.98 the result is

$$Y(\omega) = H(\omega)X(\omega).$$

(3.99)

which will be used in next to relate specific input-output for fatigue crack growth under the thermal spectrum.

The frequency response function $H(\omega)$ is useful in relating input and output power spectral densities, respectively $S_X(\omega)$ with $S_Y(\omega)$

$$S_Y(\omega) = H^*(\omega) \cdot H(\omega) S_X(\omega) = \left| H(\omega) \right|^2 S_X(\omega),$$

(3.100)

where $H^*(\omega)$ is the complex conjugate of $H(\omega)$, and $S_X(\omega)$ and $S_Y(\omega)$ are the respective power spectra. Based on the Equation 3.33 the variance of output σ_y^2 may then be expressed as

$$\sigma_Y^2 = \int_{-\infty}^{\infty} S_Y(\omega)d\omega = \int_{-\infty}^{\infty} \left| H(\omega) \right|^2 S_X(\omega)d\omega.$$

(3.101)

The relation between input and output means may be given in terms of the frequency response function for zero frequencies.

3.4.2 The Sinusoidal Frequency Response

The frequency response $H(\omega)$, which is a complex function, may be expressed also in the polar form as

$$H(\omega) = \left| H(\omega) \right| e^{\phi(i\omega)},$$

(3.102)

with its complex conjugate

$$H^*(\omega) = \left| H(\omega) \right| e^{-\phi(i\omega)}.$$

(3.103)

Let consider the steady-state response of a linear single-input, single-output system (SISO) [6] to a sinusoidal input of the form

$$u(t) = A\sin(\omega t + \psi),$$

(3.104)

where A is the amplitude of the input and ψ is an arbitrary phase angle. With Euler's formulas which relate complex exponential to sinusoidal and cosinusoidal waveforms as

$$\sin(\omega t) = \frac{1}{2i}\left(e^{i\omega t} - e^{-i\omega t}\right), \tag{3.105}$$

$$\cos(\omega t) = \frac{1}{2i}\left(e^{i\omega t} + e^{-i\omega t}\right), \tag{3.106}$$

the input may be written as

$$u(t) = A\sin(\omega t + \psi) = \frac{A}{2i}\left(e^{i(\omega t+\psi)} - e^{-i(\omega t+\psi)}\right). \tag{3.107}$$

Equation (3.107) shows that the real input $u(t)$ can be expressed as the sum of two complex exponential components

$$u_1(t) = \frac{A}{2i}e^{i(\omega t+\psi)}, \tag{3.108}$$

and

$$u_2(t) = -\frac{A}{2i}e^{-i(\omega t+\psi)}. \tag{3.109}$$

The principle of superposition allows the sinusoidal response to be written as the sum of the responses to the two complex exponential components:

$$y_s(t) = y_{s1}(t) + y_{s2}(t) = \frac{A}{2i}H(i\omega)e^{i(\omega t+\psi)} - \frac{A}{2i}H(-i\omega)e^{-i(\omega t+\psi)}. \tag{3.110}$$

If $H(i\omega)$ is written in its polar form, and $H(-i\omega)$ is described by Equation

$$H(-i\omega) = |H(i\omega)|e^{-\phi(i\omega)}, \tag{3.111}$$

then $y_s(t)$ becomes

$$
\begin{aligned}
y_s(t) &= \frac{A}{2i}|H(\omega)|\left(e^{i(\omega t+\psi)}e^{i\phi(i\omega)} - e^{-i\omega t}e^{-i(\phi(i\omega)+\psi)}\right) = \\
&= A|H(\omega)|\frac{1}{2i}\left(e^{i(\omega t+\psi+\phi(i\omega))} - e^{-i(\omega t+\psi+\phi(i\omega))}\right) = \\
&= A|H(\omega)|\sin\left(\omega t + \psi + \phi(i\omega)\right)
\end{aligned}
\tag{3.112}
$$

The steady-state sinusoidal response is a sinusoidal function of the same angular frequency ω as the input, but modified in its amplitude by the factor $|H(\omega)|$, and shifted in phase by the quantity $\phi(i\omega)$. Thus, in general, the steady-state response of a linear single-input, single-output system to a sinusoidal input

$u(t)=A\ sin\,\omega t$ can be characterized in terms of the magnitude of the frequency response function $\left|H(\omega)\right|$, and the phase shift $\phi(i\omega)=\angle H(\omega)$.

The magnitude of the frequency response represents the ratio of the output amplitude to the input amplitude as a function of frequency.

References

[1] Papoulis, A.: Probability, Random Variables, and Stochastic Processes. McGraw-Hill, New York (1984)

[2] Poularikas, A.D.: Probability and Stochastic Processes. In: The Handbook of Formulas and Tables for Signal Processi. CRC Press LLC (1999)

[3] Yung-Li, L., et al.: Fatigue testing and analysis (Theory and Practice), Elsevier Butterworth–Heinemann (2005)

[4] Rice, S.O.: Mathematical Analysis of Random noise. In: Selected papers on Noise and Stochastic Processes, Dover, New-York, pp. 139–294 (1954)

[5] Heller, R.A., Thangjitham, S.: Probabilistic Methods in Thermal Stress Analysis. In: Hetnarski, R.B. (ed.) Thermal Stress II, Elsevier Science Publishers B.V (1987)

[6] Vaseghi, S.V.: Advanced digital signal processing and noise reduction. John Wiley & Sons Ltd (2006)

Chapter 4
Stochastic Model for Thermal Fatigue Crack Growth

Abstract. In this chapter, the stochastic mathematical model is developed for the thermal fatigue crack growth phenomenon in the metallic pipe of a structural component. Any stochastic fatigue crack growth model used in time-reliability analysis must include a part means for incorporating randomness in service loads, and also another one, which should include a description of statistical characteristics of crack growth under constant amplitude loadings. Time–dependent fluctuation of temperature should be correlated with time dependent fluctuation of crack growth from the deterministic crack growth law. The number of loading cycles is a discrete variable with respect to time. When time-dependent stochastic analysis is conducted, the number of loading cycles is modified into a continuous variable by introducing an average cyclic rate. By stochastic analysis of a stationary Gaussian narrow-band process, we deal with the expected value of crack growth rate and expected rate of peak crossing (or mean rate of maxima) as well, in order to assess the thermal fatigue crack growth. The uncertainties in initial crack depth and Paris's law constants will be accounted by Monte Carlo simulation based on a properly limit state function (damage criterion). A method of crack growth assessment of linear elastic fracture mechanics (LEFM) for a stationary Gaussian narrow-band temperature fluctuation is given in this book. The model of stochastic fatigue crack growth is developed for cylindrical geometry, for which analytical solutions for temperature and associated elastic stresses were obtained in previous work. For the stochastic approach of crack growth due to random thermal fluctuations, only temporal incoherence is accounted and not any degree of spatial coherence has been taken into account.

4.1 The Main Steps of the Modeling

The parameters that affect structural fatigue performance include applied stress, geometry of structural details, properties of the material, and operating environment. A widely accepted empirical crack growth law originally was suggested by Paris, Gomez and Abderson (1961) [1]

© Springer International Publishing Switzerland 2015

V. Radu, *Stochastic Modeling of Thermal Fatigue Crack Growth,*
Applied Condition Monitoring 1, DOI: 10.1007/978-3-319-12877-1_4

$$\frac{da}{dN} = C\left(\Delta K\right)^{n} \tag{4.1}$$

where da/dN is the increment of fatigue crack advance per cycle, and ΔK is the range of the stress intensity factor, which is related in linear elastic fracture mechanics to far-field nominal stress range, and component geometry factor. C and n are empirical constants dependent on material property and the environment. The regression analysis leading to Equation (4.1) only describes the crack growth rate in the median sense. Investigation of the randomness of fatigue crack growth rate under service load conditions must consider the statistical characteristics of crack growth law under constant amplitude loadings and the randomness of loadings that gives rise to fatigue under variable amplitude loads.

Any stochastic fatigue crack growth model used in time-reliability analysis must include a part means for incorporating randomness in service loads, and also another one, which should include a description of statistical characteristics of crack growth under constant amplitude loadings.

The selection of an appropriate stochastic model for fatigue crack growth depends on the nature of uncertainty to be interpreted. The application of stochastic fatigue analysis in this study is focused on the thermal fatigue damaging phenomenon in mixing tees of nuclear components. Time–dependent fluctuation of temperature should be correlated with time dependent fluctuation of crack growth from the deterministic crack growth law. The number of loading cycles is a discrete variable with respect to time. When time-dependent stochastic analysis is conducted, the number of loading cycles is modified into a continuous variable by introducing an average cyclic rate. In this case, we have [2]

$$\frac{dA}{dt} = \frac{dA}{dN}\frac{dN}{dt} = \nu_{p}\frac{dA}{dN}, \tag{4.2}$$

in which A is the random flaw size (uppercase case denote random variables or processes) and ν_{p} is the mean rate of maxima, that is constant for a stationary random process.

By means of stochastic analysis of a stationary Gaussian narrow-band process, we deal with the expected value of crack growth rate and expected rate of peak crossing (or mean rate of maxima) as well, in order to apply Equation (4.2) for thermal fatigue crack growth. The uncertainties in initial crack depth and Paris law constants will be accounted by Monte Carlo simulation based on a properly limit state function (damage criterion).

A method of crack growth assessment of linear elastic fracture mechanics (LEFM) for a stationary Gaussian narrow-band temperature fluctuation is given in this book. The model of stochastic fatigue crack growth is developed for cylindrical geometry, for which analytical solutions for temperature and associated elastic stresses have been obtained in previous works [7, 8, 9, 10, 11]. For the stochastic approach of crack growth due to random thermal fluctuations, only temporal incoherence is accounted and not any degree of spatial coherence has been taken into account [3], [4], [5], [6].

The main steps of the procedure are given below:

a) The sinusoidal thermal loading is applied on the inner pipe surface and the external one is adiabatic; the solution of temperature distribution through thickness for an arbitrary frequency is obtained;

b) Approximation of frequency, temperature response function is made based on the general frequency output in case of sinusoidal input, and also by keeping an appropriate conservatism its magnitude is established;

c) By using the linear elasticity theory for hollow cylinder, the analytical solutions for thermal stress components are derived (i.e. radial, hoop and axial) for various boundary conditions;

d) The magnitude of the frequency response function for thermal stresses is obtained, by using the corresponding frequency response for temperature;

e) Calculating the stress intensity factor (SIF) frequency response is based on the polynomial fitting of stress response profile through thickness, which is usually used in the case of a very high-stress gradient in the pipe wall;

f) The power spectral density of SIF is inferred from the power spectral density of the thermal spectrum applied on the inner pipe surface, by considering the stationary Gaussian narrow-band temperature spectrum;

g) The main moments of PSD for SIF are calculated (mean up crossing rate and amplitude distribution);

h) The crack propagation rate is converted from respect to cycles into time variable respect, and the expected value of crack growth rate is obtained by assuming linear summation of damage and ignore the effect of positive maxima in the crack propagation law;

i) A probabilistic input for Paris's law integration is used to account the variability in initial crack depth of flaws and also in C scaling parameter;

j) The lifetime for crack growth due to thermal fluctuations is given in the form of probability of failure (or index reliability), and its assessment is based on the limit state function with Monte Carlo simulation.

Note that the calculations are limited to LEFM, and any plasticity or retardation effects are ignored. The crack initiation is a separately subject, and it is no longer accounted here; in this respect the crack propagation, that follows after the initiation period is described only by probabilistic distribution function of initial crack depth of flaws. The stress-free temperature that influences the SIF magnitude during its fluctuations is not considered as well.

4.2 Statistical Properties of the Thermal Spectrum

The main assumption is that the temperature spectrum at the inner pipe surface can be modeled as a stationary Gaussian narrow-band process, and its power spectral density is known. Firstly, an analytical solution for temperature distribution in the wall-thickness of the pipe is derived under sinusoidal thermal loading at the inner surface. In the next step, the frequency response function is proposed.

4.2.1 Analytical Solution of Temperature Distribution under Sinusoidal Thermal Loading

The analytical solution for time dependent temperature profile for an infinite hollow cylinder has been developed in [7]. A short overview will be given in the follow.

Assuming an infinite hollow cylinder made of a homogeneous isotropic material, with inner and outer radii $r_i=a$ and $r_o=b$, the 1D heat diffusion equation has the form

$$\frac{\partial^2 \Theta}{\partial r^2} + \frac{1}{r}\frac{\partial \Theta}{\partial r} = \frac{1}{k}\frac{\partial \Theta}{\partial t}, \ (a \leq r \leq b, \ t \geq 0), \tag{4.3}$$

where

$$\Theta(r,t) = T(r,t) - T_0, \tag{4.4}$$

is the temperature change from the reference temperature at any radial position r and at time t.

The reference temperature T_0 is the body temperature in the unstrained state or the ambient temperature before changing of temperature. The thermal diffusivity is defined as

$$k = \frac{\lambda}{\rho c}, \tag{4.5}$$

with λ is the thermal conductivity, ρ is the mass density and c is the specific heat conduction. The solution $\Theta(r,\ t)$ must satisfy the boundary conditions

$$\Theta(a,t) = q(t), \ (t \geq 0), \tag{4.6}$$

$$\Theta(b,t) = 0, \ (t \geq 0), \tag{4.7}$$

and the initial condition

$$\Theta(r,0) = 0, \ (a \leq r \leq b). \tag{4.8}$$

The function $q(t)$ is a known function of time representing the thermal boundary condition applied at the inner surface of the cylinder. By using the finite Hankel transform, in the way presented in Appendix A; we are able to obtain the analytical solution for arbitrary boundary condition, $q(t)$. The differential Equation (4.3) contains a linear operator $\mathbf{L}$, applied to a function $\Theta(r,t)$ in general form of

Equation (A-9) from Appendix A, with $v=0$. Following Equation (A-14) is introduced the transform $\overline{\Theta}(s_n,t)$ as follows

$$\overline{\Theta}(s_n,t) = \mathrm{H}\left[\Theta(r,t); s_n\right], \tag{4.9}$$

and make use the result Equation (A-17) to show that

$$\mathrm{H}\left[\frac{\partial^2 \Theta}{\partial r^2} + \frac{1}{r}\frac{\partial \Theta}{\partial r}; s_n\right] = \frac{2J_0(s_n a)}{\pi J_0(s_n a)}\Theta(b,t) - \frac{2}{\pi}\Theta(a,t) - s_n^2 \overline{\Theta}(s_n,t) \tag{4.10}$$

With boundary conditions from Equations (4.6) and (4.7), $\overline{\Theta}(s_n,t)$ satisfies the differential equation

$$\left(\frac{\partial}{\partial t} + ks_n^2\right)\overline{\Theta}(s_n,t) = -\frac{2}{\pi}q(t), \tag{4.11}$$

and the initial condition

$$\overline{\Theta}(s_n,0) = 0. \tag{4.12}$$

Thus, the following solution for $\overline{\Theta}(s_n,t)$ is obtained

$$\overline{\Theta}(s_n,t) = -\frac{2k}{\pi}e^{-ks_n^2 t}\int_0^t q(\tau)d\tau. \tag{4.13}$$

The solution $\Theta(r,t)$ may be obtained by the inversion theorem for operator H as in Equation (A-16)

$$\Theta(r,t) \equiv \mathrm{H}^{-1}\left[\overline{\Theta}(s_n,t); r\right] = \frac{\pi^2}{2}\sum_{n=1}^{\infty}\frac{s_n^2 J_0^2(s_n b)\overline{\Theta}(s_n,t)}{J_0^2(s_n a) - J_0^2(s_n b)}\left[J_0(s_n r)Y_0(s_n a) - J_0(s_n a)Y_0(s_n r)\right]. \tag{4.14}$$

Here $J_0(z)$, $Y_0(z)$ are Bessel's functions of first and second kind of order 0. Finally, the temperature distribution in the thickness is given by analytical solution:

$$\Theta(r,t) = k\pi\sum_{n=1}^{\infty}\frac{s_n^2 J_0^2(s_n b)}{J_0^2(s_n b) - J_0^2(s_n a)}\left[J_0(s_n r)Y_0(s_n a) - J_0(s_n a)Y_0(s_n r)\right]\left[e^{-ks_n^2 t}\int_0^t e^{ks_n^2 \tau}q(\tau)d\tau\right] \tag{4.15}$$

where s_n are the positive roots of the transcendental equation

$$Y_0(s_n a)J_0(s_n b) - J_0(s_n a)Y_0(s_n b) = 0. \tag{4.16}$$

The analytical solution for temperature distribution from Equation (4.15): may be written as follows:

$$\Theta(r,t) = k\pi \sum_{n=1}^{\infty} \Theta_1(a,b,s_n) \cdot \Theta_2(a,r,s_n) \cdot \Theta_3(t,s_n), \qquad (4.17)$$

where

$$\Theta_1(a,b,s_n) = \frac{s_n^2 J_0^2(s_n b)}{J_0^2(s_n b) - J_0^2(s_n a)}, \qquad (4.18)$$

$$\Theta_2(a,r,s_n) = Y_0(s_n a) J_0(s_n r) - J_0(s_n a) Y_0(s_n r), \qquad (4.19)$$

$$\Theta_3(t,s_n) = e^{-ks_n^2 t} \int_0^t e^{ks_n^2 \tau} q(\tau) d\tau. \qquad (4.20)$$

To evaluate the time-dependent term $\Theta_3(t, s_n)$, it is assumed the sinusoidal thermal loading

$$q(t) = \Theta_0 \cdot \sin(\omega t) = \Theta_0 \cdot \sin(2\pi f t), \qquad (4.21)$$

where Θ_0 is the amplitude of temperature wave, ω and f corresponds to angular frequency in radians/second and cycles/second (Hz), respectively, and t is time. By substituting the Equation
(4.21) and performing the integral, the result is

$$\Theta_3(\omega,t,s_n) = \Theta_0 \frac{\omega e^{-ks_n^2 t} + \left(ks_n^2\right)\sin(\omega t) - \omega\cos(\omega t)}{\left(ks_n^2\right)^2 + \omega^2}. \qquad (4.22)$$

By inserting the Equations (4.18), (4.19) and (4.20) into Equation (4.17), the final form of analytical solution for temperature distribution through wall-thickness of the pipe is

$$\Theta(r,\omega,t) = k\pi \sum_{n=1}^{\infty} \frac{s_n^2 J_0^2(s_n b)}{J_0^2(s_n b) - J_0^2(s_n a)} \left[J_0(s_n r) Y_0(s_n a) - J_0(s_n a) Y_0(s_n r) \right] \times$$

$$\times \left[\Theta_0 \frac{\omega e^{-ks_n^2 t} + \left(ks_n^2\right)\sin(\omega t) - \omega\sin(\omega t)}{\left(ks_n^2\right)^2 + \omega^2} \right] \qquad (4.23)$$

In the paper [7] the predictions of the analytical solution, given by Equation (4.23), have been checked by finite-element analyses performed with ABAQUS computer code (Figure 4.1), with good agreement [7,8].

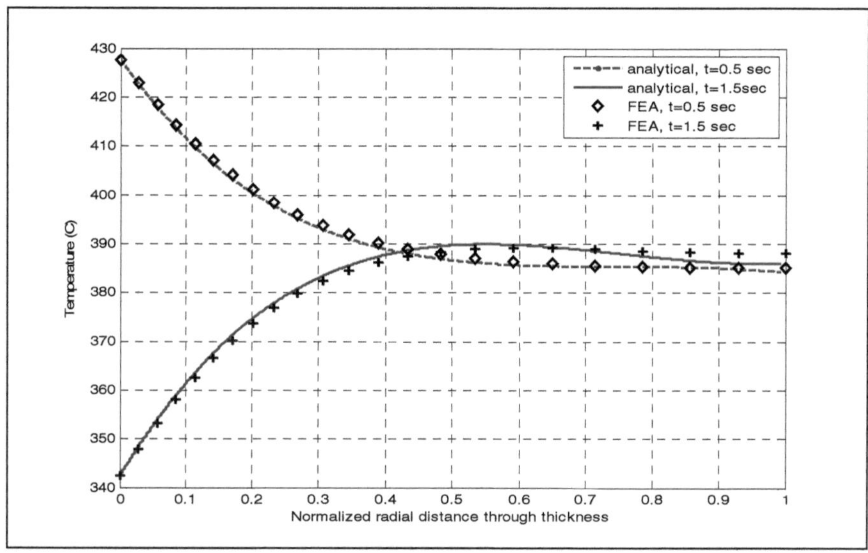

Fig. 4.1 Comparison between analytical prediction for temperature and FEA for a pipe striped at the inner surface with sinusoidal thermal loading at frequency f=0. 5Hz [7]

4.2.2 Approximation of the Temperature Frequency Response Function for Hollow Cylinder

From practical point of view it is necessary to obtain the magnitude of frequency response function for temperature and associated thermal stresses by considering a certain geometry, spectrum loading and material properties. This way will give us the flexibility to manipulate the analytical expressions from the general form to specific form, most suitable for application itself. Therefore, in this book, we have chosen a well-known damaging case study for application of stochastic methodology, namely Civaux case study [9, 10,11]. In the appendix, B is given a short description provided by reference [12]. The use of the analytical solution from Equation (4.23) needs the positive roots s_n of transcendental Equation (4.16). Previous work [8] showed that by using first one hundred positive roots provides a stable and optimized analytical response for temperature. For Civaux case, where the geometry of the pipe consists in inner and outer radii by $r_i=a=0.120$m and $r_o=b=0.129$ m, the first 100 roots of transcendental Equation (4.16) are given in Appendix C.

The analytical solution for temperature variation through wall-thickness from Equation 4.23, due to sinusoidal loading at the inner pipe surface (Equation 4.21), must be adapted to a suitable form that allows the extraction of temperature frequency response magnitude. In the next, we work on the Equation 4.22, which contains the variable frequency, ω. A first attempt is to neglect the exponential term from Equation (4.22)

$$\Theta_3(\omega,t,s_n) = \Theta_0 \frac{\omega e^{-ks_n^2 t} + \left(ks_n^2\right)\sin\left(\omega t\right) - \omega\cos\left(\omega t\right)}{\left(ks_n^2\right)^2 + \omega^2} \simeq \Theta_0 \frac{\left(ks_n^2\right)\sin\left(\omega t\right) - \omega\cos\left(\omega t\right)}{\left(ks_n^2\right)^2 + \omega^2}. \quad (4.24)$$

When we perform this operation, we have to check its availability to predict the temperature fluctuations inside of the pipe-wall. By inserting Equation 4.24 into Equation 4.23, the first approximation o temperature response is

$$\Theta\left(r,\omega,t\right) \simeq k\pi \sum_{n=1}^{\infty} \frac{s_n^2 J_0^2\left(s_n b\right)}{J_0^2\left(s_n b\right) - J_0^2\left(s_n a\right)} \left[J_0\left(s_n r\right) Y_0\left(s_n a\right) - J_0\left(s_n a\right) Y_0\left(s_n r\right) \right] \times$$
$$\times \left[\Theta_0 \frac{\left(ks_n^2\right)\sin\left(\omega t\right) - \omega\sin\left(\omega t\right)}{\left(ks_n^2\right)^2 + \omega^2} \right] \tag{4.25}$$

For the Civaux case, we consider [10, 11] the input signal (sinusoidal thermal loading) at the inner surface as

$$q(t) = \Theta_0 \cdot \sin\left(\omega t\right) = \Theta_0 \cdot \sin\left(2\pi f t\right), \tag{4.26}$$

with the amplitude of Θ_0=60°C. The frequencies of loading will be in range of 0.1 to 1.0 Hz, where the critical frequencies for crack growth are placed [11].

The results given by Equations 4.23 and 4.25 are compared in Figure 4.2.

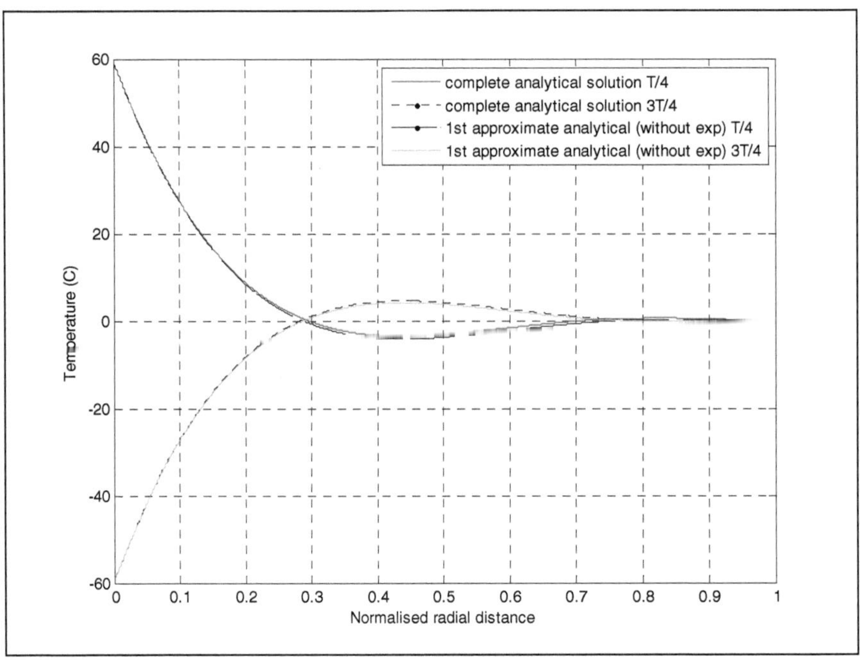

Fig. 4.2 First approximate of analytical predictions for temperature compared with complete analytical solution through wall-thickness of a pipe striped at inner surface with sinusoidal thermal loading at frequency f=0.5Hz

As it can be seen, the both predictions have, practically, the same temperature profiles through thickness, in the case of Civaux geometry, which means that the influence of the exponential term can be neglected.

In a second step, it is important to manipulate the last term of Equation (4.25) to obtain the general form of response frequency response similar to Equation 3.112. Thus

$$\Theta_3(\omega,t,s_n) \simeq \Theta_0 \frac{\left(ks_n^2\right)\sin\left(\omega t\right)-\omega\cos\left(\omega t\right)}{\left(ks_n^2\right)^2+\omega^2} = \Theta_0\left(ks_n^2\right)\frac{\sin\omega t-\left(\dfrac{\omega}{ks_n^2}\right)\cos\omega t}{\left(ks_n^2\right)^2+\omega^2}. \qquad (4.27)$$

Let consider

$$\tan\phi\left(\omega,s_n\right) = \frac{\omega}{Ks_n^2}, \qquad (4.28)$$

then Equation (4.27) becomes

$$\Theta_3(\omega,t,s_n) \simeq \Theta_0\left(ks_n^2\right)\frac{\sin\omega t-\left(\dfrac{\omega}{ks_n^2}\right)\cos\omega t}{\left(ks_n^2\right)^2+\omega^2} = \Theta_0\frac{\sin\left[\omega t-\tan^{-1}\left(\dfrac{\omega}{ks_n^2}\right)\right]}{\sqrt{\left(ks_n^2\right)^2+\omega^2}} =$$

$$= \Theta_0\frac{\sin\left[\omega t-\varphi(\omega,s_n)\right]}{\sqrt{\left(ks_n^2\right)^2+\omega^2}} \qquad (4.29)$$

With 4.29, Equation 4.25 gives the first approximation in the following form

$$\Theta(r,\omega,t) \simeq \sum_{n=1}^{\infty} \frac{k\pi\dfrac{s_n^2 J_0^2\left(s_n b\right)}{J_0^2\left(s_n b\right)-J_0^2\left(s_n a\right)}\left[J_0\left(s_n r\right)Y_0\left(s_n a\right)-J_0\left(s_n a\right)Y_0\left(s_n r\right)\right]}{\sqrt{\left(ks_n^2\right)^2+\omega^2}}\times$$

$$\times\Theta_0\sin\left[\omega t-\varphi\left(\omega,s_n^2\right)\right] \qquad (4.30)$$

Each term of sum from Equation (4.30) may be expressed as

$$T_n\left(r,\omega,t\right) \simeq \Theta_0\frac{k\pi\dfrac{s_n^2 J_0^2\left(s_n b\right)}{J_0^2\left(s_n b\right)-J_0^2\left(s_n a\right)}\left[J_0\left(s_n r\right)Y_0\left(s_n a\right)-J_0\left(s_n a\right)Y_0\left(s_n r\right)\right]}{\sqrt{\left(ks_n^2\right)^2+\omega^2}}\times$$

$$\times\sin\left[\omega t-\varphi\left(\omega,s_n^2\right)\right] \qquad (4.31)$$

which is in the form of Equation (3.112).

Because we interested only in the magnitude of frequency temperature response, the Equation (4.31) may be written in the general form as

$$T_n\left(r,\omega,t,s_n\right) \simeq \Theta_0\left|H_{T,n}\left(\omega,r,s_n\right)\right|\times\sin\left[\omega t-\varphi\left(\omega,s_n^2\right)\right], \qquad (4.32)$$

with

$$\left| H_{T,n}\left(r,\omega,s_n \right) \right| = abs\left\{ \frac{k\pi \dfrac{s_n^2 J_0^2\left(s_n b \right)}{J_0^2\left(s_n b \right) - J_0^2\left(s_n a \right)}\left[J_0\left(s_n r \right) Y_0\left(s_n a \right) - J_0\left(s_n a \right) Y_0\left(s_n r \right) \right]}{\sqrt{\left(k s_n^2 \right)^2 + \omega^2}} \right\}. \qquad (4.33)$$

At this point, to force the temperature response from Equation 4.25 to follow a typical frequency response similar to (4.32) we shall consider Equation (4.32), in a second approximation, with phase $\varphi\left(\omega, s_n \right)$ being constant. The result is then

$$\Theta\left(r,\omega,t \right) \simeq \Theta_0\left[\sum_{n=1}^{\infty} \left| H_{T,n}\left(r,\omega,s_n \right) \right| \right] \times \sin\left[\omega t - \varphi \right]. \qquad (4.34)$$

Finally, we designate the magnitude of temperature frequency response as

$$\left| H_T\left(r,\omega \right) \right| = \sum_{n=1}^{\infty} abs\left\{ \frac{k\pi \dfrac{s_n^2 J_0^2\left(s_n b \right)}{J_0^2\left(s_n b \right) - J_0^2\left(s_n a \right)}\left[J_0\left(s_n r \right) Y_0\left(s_n a \right) - J_0\left(s_n a \right) Y_0\left(s_n r \right) \right]}{\sqrt{\left(k s_n^2 \right)^2 + \omega^2}} \right\} \qquad (4.35)$$

and temperature fluctuation in the pipe-wall is given by

$$\Theta\left(r,\omega,t \right) \simeq \Theta_0 \left| H_T\left(r,\omega \right) \right| \times \sin\left[\omega t - \varphi \right]. \qquad (4.36)$$

In a conservative way, if the lag phase is approximate as $\varphi \approx 0$, a comparison between temperature profiles through thickness predicted by Equation (4.23) - or (4.25) - and Equation (4.36) with (4.35) is displayed in Figure 4.3.

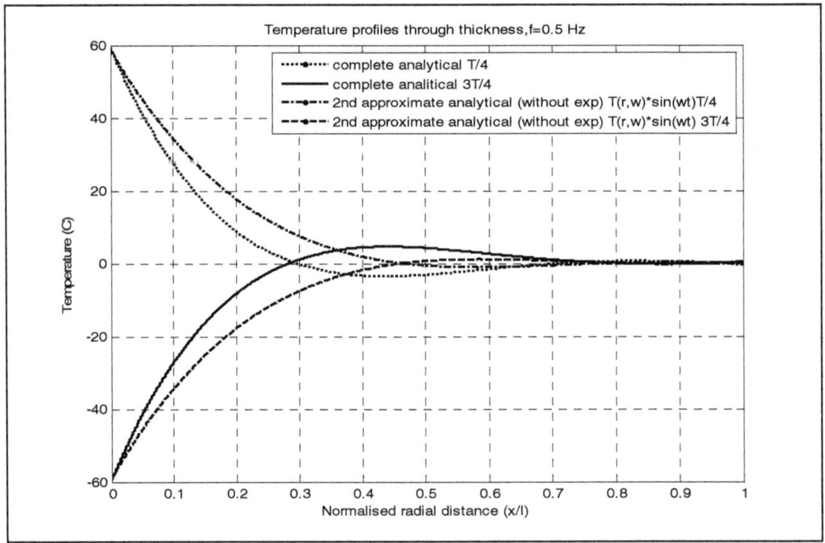

Fig. 4.3 Comparison between predictions of temperature profile from complete analytical solution and those obtained by means of analytical temperature frequency response function in the pipe wall

With the same temperature range at the inner pipe surface and with a slightly deeper penetration, the prediction of temperature response with assumed magnitude of frequency temperature response given by Equation 4.36 may be reasonable accounted. Figure 4.4 shows the influence of loading frequency, in case of sinusoidal input, on the temperature frequency response magnitude, given by Equation 4.35, for several points inside of the wall.

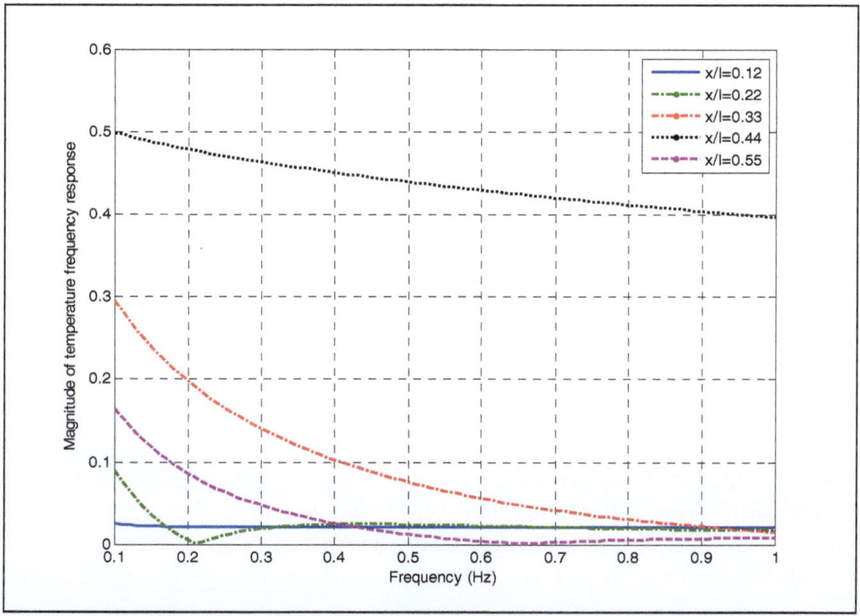

Fig. 4.4 Dependence of temperature frequency response magnitude on loading frequency for various depths through thickness (l is wall-thickness and x originates at the inner pipe surface)

The highest value of response is obtained at $x/l=4/9=0.44$, in the Civaux case geometry, and for deeper points located inside of the pipe-wall values go down for the whole range of frequencies.

4.3 Modeling of the Stress Response to Random Thermal Input

To obtain the stress frequency response function a similar approach as in the previous chapter is used. The general solution of elastic thermal stress components (hoop, radial and axial) due to a sinusoidal loading at the inner surface is, firstly, derived. Subsequently, the stress frequency response is obtained by means of temperature frequency response function, which will be used in the corresponding analytical solution. A comparison with FEA prediction is made and a sensitivity analysis versus frequency range as well.

4.3.1 Analytical Solutions for Elastic Thermal Stress due to Sinusoidal Thermal Input

For thermal stress evaluation, we assumed that the thermo-mechanical properties are not changing during the thermal transient analyses.

The one-dimensional equilibrium equation in the radial direction for a hollow cylinder is [7, 8,13]:

$$\frac{d\sigma_r}{dr} + \frac{\sigma_r - \sigma_\theta}{r} = 0,$$

(4.37)

where σ_r and σ_θ are the radial and hoop stresses respectively. In the axis symmetric problem with small strains, the strain-displacement relations are:

$$\varepsilon_r = \frac{du}{dr},$$

(4.38)

$$\varepsilon_\theta = \frac{u}{r},$$

(4.39)

$$\varepsilon_{r\theta} = 0.$$

(4.40)

where u is the radial displacement.

The displacement technique has been used to solve the axis symmetric problems of the hollow cylinder. The components of stress in cylindrical coordinates can be expressed as

$$\sigma_r = \frac{E'}{1-v'^2}\left[\frac{du}{dr} + v'\cdot\frac{u}{r} - (1+v')\cdot\alpha'\cdot\Theta + (1+v')\cdot c'\right],$$

(4.41)

$$\sigma_\theta = \frac{E'}{1-v'^2}\left[v'\cdot\frac{du}{dr} + \frac{u}{r} - (1+v')\cdot\alpha'\cdot\Theta + (1+v')\cdot c'\right],$$

(4.42)

$$\sigma_{r\theta} = 0.$$

(4.43)

In the case of plane strain and plane stress, the meanings of the constants from Equations (4.41) to (4.43) are:

$$E' = \begin{cases} \dfrac{E}{1-v^2} & \text{E–Young modulus,} \\ E \end{cases}$$

(4.44)

$$v' = \begin{cases} \dfrac{v}{1-v} \\ v \end{cases} \quad v \text{ Poisson's ratio,} \tag{4.45}$$

$$\alpha' = \begin{cases} (1+v)\alpha \\ \alpha \end{cases} \quad \alpha \text{ -coefficient of the linear thermal expansion,} \tag{4.46}$$

$$c' = \begin{cases} v\varepsilon_0 \\ 0 \end{cases} \quad \varepsilon_0 \text{ - constant axial strain for plane strain state.} \tag{4.47}$$

The substitution of Equations (4.41) and (4.42) in Equation (4.37) yields

$$\frac{d}{dr}\left[\frac{1}{r}\cdot\frac{d(r\cdot u)}{dr}\right] = (1+v')\cdot\alpha'\cdot\frac{d\Theta(r,t)}{dr}. \tag{4.48}$$

The general solution of Equation (4.48) is

$$u = (1+v')\cdot\alpha'\cdot\frac{1}{r}\int_r \theta(r,t)\cdot r\cdot dr + C_1\cdot r + \frac{C_2}{r}. \tag{4.49}$$

The integration constants C_1 and C_2 may be determined from the boundary conditions. The radial stress component is negligible for thin-walled cylinder compared to the hoop and axial stress components. The hoop and axial stress components are given in the following relationships in the case of plane strain [13]:

$$\sigma_\theta(r,\omega,t) = \frac{\alpha\cdot E}{1-v}\left[\frac{1}{r^2}\cdot I_1(r,\omega,t) + \frac{r^2+a^2}{r^2\cdot(b^2-a^2)}\cdot I_2(\omega,t) - \Theta(r,\omega,t)\right], \tag{4.50}$$

$$\sigma_z(r,\omega,t) = \frac{\alpha\cdot E}{1-v}\left[\frac{2v}{b^2-a^2}\cdot I_2(\omega,t) - \Theta(r,\omega,t)\right] \text{ for } \varepsilon_z=0, \tag{4.51}$$

$$\sigma_z(r,\omega,t) = \frac{\alpha\cdot E}{1-v}\left[\frac{2}{b^2-a^2}\cdot I_2(\omega,t) - \Theta(r,\omega,t)\right] \text{ for } \varepsilon_z=\varepsilon_0. \tag{4.52}$$

The mathematical relationships for $I_1(r,\omega t)$ and $I_2(\omega t)$ are given in [7, 8]. Because we need to work on the temperature part, here are the complete expressions for the case of sinusoidal thermal loading at the inner surface:

$$I_1(r,\omega,t) = \int_a^r \Theta(r,t) \cdot r \cdot dr = k \cdot \pi \cdot \sum_{n=1}^{\infty} \frac{s_n^2 \cdot J_0^2(s_n \cdot b)}{J_0^2(s_n \cdot b) - J_0^2(s_n \cdot a)} \times$$

$$\times \left[\frac{1}{s_n} \left\{ Y_o(s_n \cdot a) \cdot [r \cdot J_1(s_n \cdot r) - a \cdot J_1(s_n \cdot a)] - J_o(s_n \cdot a) \cdot [r \cdot Y_1(s_n \cdot r) - a \cdot Y_1(s_n \cdot a)] \right\} \right] \times$$

$$\times \left[\Theta_0 \cdot \frac{\omega \cdot e^{-k \cdot s_n^2 \cdot t} + (k \cdot s_n^2) \cdot \sin(\omega \cdot t) - \omega \cdot \cos(\omega \cdot t)}{(k \cdot s_n^2)^2 + \omega^2} \right] \qquad (4.53)$$

$$I_2(\omega,t) = \int_a^b \Theta(r,t) \cdot r \cdot dr = k \cdot \pi \cdot \sum_{n=1}^{\infty} \frac{s_n^2 \cdot J_0^2(s_n \cdot b)}{J_0^2(s_n \cdot b) - J_0^2(s_n \cdot a)} \times$$

$$\times \left[\frac{1}{s_n} \left\{ Y_o(s_n \cdot a) \cdot [b \cdot J_1(s_n \cdot b) - a \cdot J_1(s_n \cdot a)] - J_o(s_n \cdot a) \cdot [b \cdot Y_1(s_n \cdot b) - a \cdot Y_1(s_n \cdot a)] \right\} \right] \times$$

$$\times \left[\Theta_0 \cdot \frac{\omega \cdot e^{-k \cdot s_n^2 \cdot t} + (k \cdot s_n^2) \cdot \sin(\omega \cdot t) - \omega \cdot \cos(\omega \cdot t)}{(k \cdot s_n^2)^2 + \omega^2} \right] \qquad (4.54)$$

where s_n are the positive roots of the transcendental Equation (4.16), and $J_0(z)$, $Y_0(z)$, $J_1(z)$, $Y_2(z)$ are Bessel's functions of first and second kind of order 0 and order 1, respectively.

The complete form of Equations (4.50 to 4.52) for thermal stress components are given in Appendix D.

4.3.2 The Stress Frequency Response Function

In the next, we only work for hoop stress component, σ_θ, to find its frequency response function, but with a similar approach one can get the frequency response forms for radial (σ_r) and axial (σ_z) stress components.

The general approach is to substitute the temperature frequency response function in the solution of thermal stress components to make possible obtaining the stress frequency response function [14]. Here we start with processing the relationships for integrals $I_1(r,\omega,t)$ and $I_2(\omega,t)$ given by Equation (4.53) and (4.54) respectively. Substituting Equation (4.29) into integrals $I_1(r,\omega,t)$ and $I_2(\omega,t)$ this yields to approximate forms

$$I_1(r,\omega,t) \simeq \int_a^r \Theta(r,t) \cdot r \cdot dr = k \cdot \pi \cdot \sum_{n=1}^{\infty} \frac{s_n^2 \cdot J_0^2(s_n \cdot b)}{J_0^2(s_n \cdot b) - J_0^2(s_n \cdot a)} \times$$

$$\times \left[\frac{1}{s_n} \left\{ Y_o(s_n \cdot a) \cdot [r \cdot J_1(s_n \cdot r) - a \cdot J_1(s_n \cdot a)] - J_o(s_n \cdot a) \cdot [r \cdot Y_1(s_n \cdot r) - a \cdot Y_1(s_n \cdot a)] \right\} \right] \times$$

$$\times \left[\Theta_0 \cdot \frac{\sin\left[\omega \cdot t - \varphi(\omega, s_n) \right]}{\sqrt{\left(k \cdot s_n^2\right)^2 + \omega^2}} \right] \qquad (4.55)$$

$$I_2(\omega,t) \simeq \int_a^b \Theta(r,t)\cdot r\cdot dr = k\cdot\pi\cdot\sum_{n=1}^{\infty}\frac{s_n^2\cdot J_0^2(s_n\cdot b)}{J_0^2(s_n\cdot b)-J_0^2(s_n\cdot a)}\times$$

$$\times\left[\frac{1}{s_n}\left\{Y_o(s_n\cdot a)\cdot[b\cdot J_1(s_n\cdot b)-a\cdot J_1(s_n\cdot a)]-J_o(s_n\cdot a)\cdot[b\cdot Y_1(s_n\cdot b)-a\cdot Y_1(s_n\cdot a)]\right\}\right]\times$$

$$\times\left[\Theta_0\cdot\frac{\sin\left[\omega\cdot t-\varphi(\omega,s_n)\right]}{\sqrt{\left(k\cdot s_n^2\right)^2+\omega^2}}\right] \tag{4.56}$$

Obviously, both integrals may be written in the form similar to Equation (4.34), such as

$$I_1(r,\omega,t)\simeq\Theta_0\cdot\left[\sum_{n=1}^{\infty}\left|H_{I_1,n}(r,\omega,s_n)\right|\right]\cdot\sin(\omega t-\varphi), \tag{4.57}$$

$$I_2(\omega,t)\simeq\Theta_0\cdot\left[\sum_{n=1}^{\infty}\left|H_{I_2,n}(\omega,s_n)\right|\right]\cdot\sin(\omega t-\varphi). \tag{4.58}$$

With the temperature response from Equation (4.36) and integrals, $I_1(r,\omega,t)$ and $I_2(\omega,t)$ from Equations (4.57) and (4.58), the hoop stress from Equation (4.50) becomes

$$\sigma_\theta(r,\omega,t)\simeq\Theta_0\cdot\left|H_{\sigma_\theta}(r,\omega)\right|\cdot\sin(\omega t-\varphi). \tag{4.59}$$

Furthermore, with processing of Equation (D1) from Appendix D, the final form of magnitude for the stress frequency response is given by

$$\left|H_{\sigma_\theta}(r,\omega)\right|=abs\left\{\frac{k\cdot\pi\cdot\alpha\cdot E}{1-v}\times\left\{\left(\frac{1}{r^2}\right)\sum_{n=1}^{\infty}\frac{s_n^2\cdot J_0^2(s_n\cdot b)}{J_0^2(s_n\cdot b)-J_0^2(s_n\cdot a)}\times\right.\right.$$

$$\times\left[\frac{Y_o(s_n\cdot a)\cdot[r\cdot J_1(s_n\cdot r)-a\cdot J_1(s_n\cdot a)]-J_o(s_n\cdot a)\cdot[r\cdot Y_1(s_n\cdot r)-a\cdot Y_1(s_n\cdot a)]}{s_n\sqrt{\left(ks_n^2\right)^2+\omega^2}}\right]$$

$$+\frac{r^2+a^2}{r^2(b^2-a_i^2)}\times\sum_{n=1}^{\infty}\frac{s_n^2\cdot J_0^2(s_n\cdot b)}{J_0^2(s_n\cdot b)-J_0^2(s_n\cdot a)}\times$$

$$\times\left[\frac{Y_o(s_n\cdot a)\cdot[b\cdot J_1(s_n\cdot b)-a\cdot J_1(s_n\cdot a)]-J_o(s_n\cdot a)\cdot[b\cdot Y_1(s_n\cdot b)-a\cdot Y_1(s_n\cdot a)]}{s_n\sqrt{\left(ks_n^2\right)^2+\omega^2}}\right]- \tag{4.60}$$

$$-\sum_{n=1}^{\infty}\frac{s_n^2\cdot J_0^2(s_n\cdot b)}{J_0^2(s_n\cdot b)-J_0^2(s_n\cdot a)}\cdot\left[\frac{Y_o(s_n\cdot a)\cdot J_o(s_n\cdot r)-J_o(s_n\cdot a)\cdot Y_o(s_n\cdot r)}{s_n\sqrt{\left(ks_n^2\right)^2+\omega^2}}\right]\left.\left.\right\}\right\}$$

To see how the function for magnitude of stress frequency response works, an analysis of predictions from Equation (4.59) with (4.60), by comparison with those from FEA and complete analytical solution (Equation D1) is made. Figure 4.5 illustrates the agreement between all of them. As it can be seen from comparison, one can conclude that the magnitude of stress frequency response is reasonable described by Equation (4.59), actually in a conservative mode.

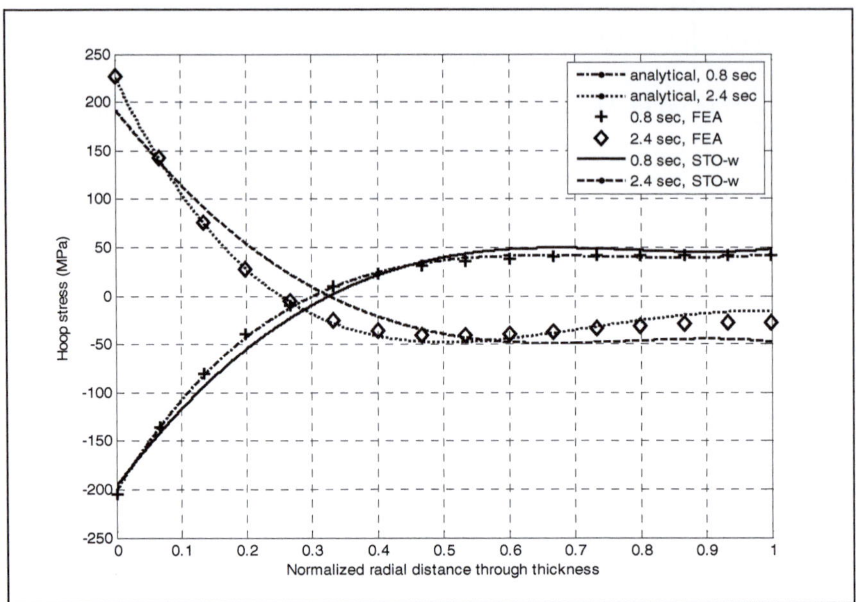

Fig. 4.5 Comparison between predictions for hoop stress: complete analytical solution, FEA, and by means of stress frequency response function (frequency of sinusoidal thermal loading f=0.3 Hz)

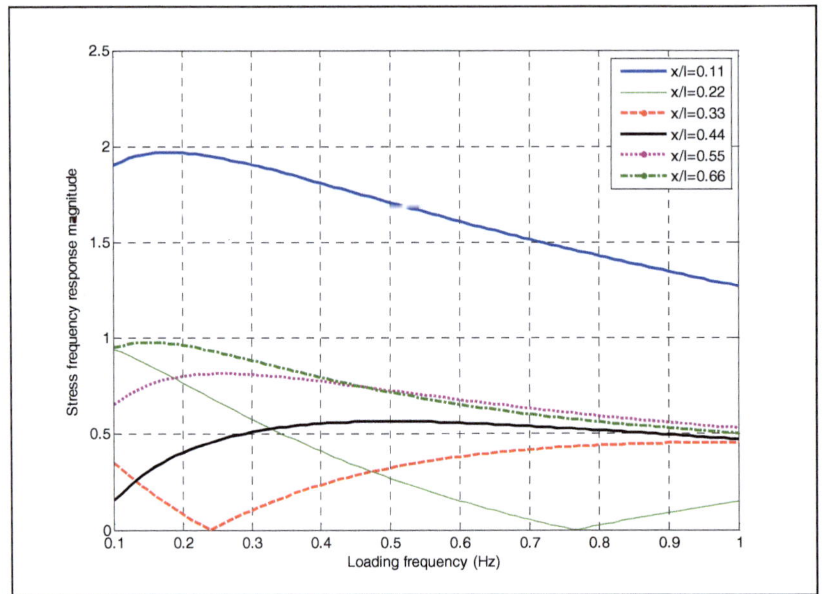

Fig. 4.6 Magnitude of stress frequency response function versus loading frequency inside of the pipe-wall

The magnitude of stress frequency response has an interesting dependence on loading frequency for points inside of the wall along the radial direction, as it is illustrated in Figure 4.6.

Note, that frequency responses have not the same dependence for temperature as those for hoop stress. Moreover, moving on into the pipe wall, for each location the frequencies for which the maxima of respective response functions is reached are not the same as well.

It is interesting to see also, how is the dependence of profiles through thickness for stress frequency response magnitude, on the loading frequency. Figure 4.7 displays profiles of function $\left|H_{\sigma_\theta}(r,\omega)\right|$, given by Equation (4.60), fitted by polynomials of 4th order, for several frequencies in the range 0.1 Hz to 1.0 Hz. This figure illustrates the sensitivity of magnitude of stress frequency response function to loading frequency, for every point of wall-thickness, and should be considered as a complement to Figure 4.6.

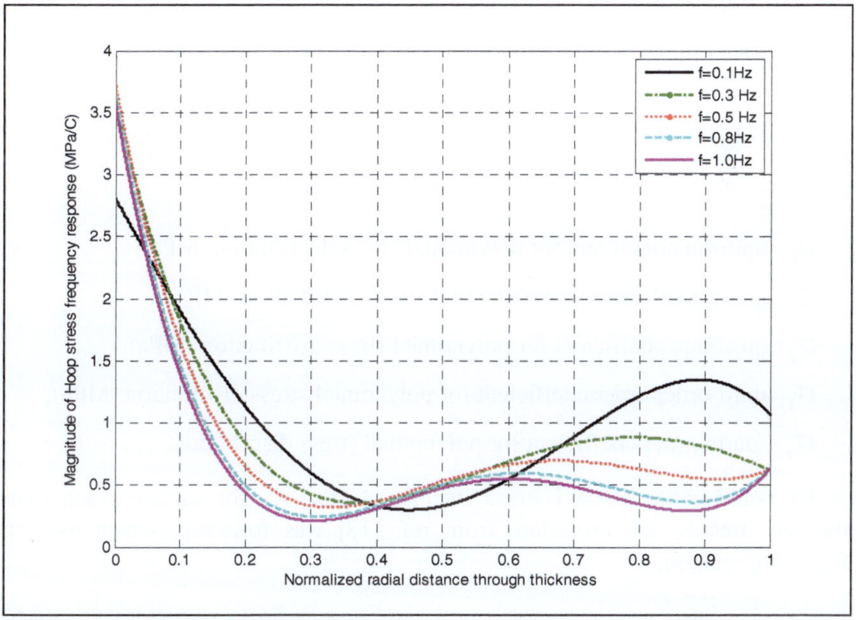

Fig. 4.7 Magnitude of stress frequency response function through wall thickness versus loading frequency inside of the pipe-wall

4.4 Stress Intensity Factor (SIF) Approach

There is assumed to be a shallow crack of infinite length on the inner surface and parallel to the tube axis. A most-used method to obtain the stress intensity factor K_I in case of through-thickness stress profile is weight-function method:

$$K(a,\omega,t) = \int_0^a \sigma_\theta(x,\omega,t) M(x)\,dx, \qquad (4.61)$$

where $M(x)$ is the weight function, and a is the crack depth.

In the case of analytical solution for hoop stress presented in the report, the solving of the corresponding integral involving Bessel's functions is impracticable.

An alternative formulation for SIF, which has been used in previous works [10, 11], is to approximate the through-thickness profile as a function of non-dimensional radial local coordinate (x/l) originating at the internal surface of the component, where l represents the thickness of the wall. For a long axial crack, our approach to derive the stress intensity factors is based on the polynomial representation of stress components through the wall-thickness of the pipe. The fourth order polynomial distribution can be used for highly non-linear stress distributions, such as the hoop stresses arising during a period of sinusoidal thermal loading, by curve-fitting the analytical stress distribution.

The general form of the fourth order polynomial distribution is [15]:

$$\sigma(x) = \sigma_0 + \sigma_1 \cdot \left(\frac{x}{l}\right) + \sigma_2 \cdot \left(\frac{x}{l}\right)^2 + \sigma_3 \cdot \left(\frac{x}{l}\right)^3 + \sigma_4 \cdot \left(\frac{x}{l}\right)^4, \qquad (4.62)$$

where:

σ_0 -uniform coefficient for polynomial stress distribution (MPa);

σ_1 -linear coefficient for polynomial stress distribution (MPa);

σ_2 -quadratic coefficient for polynomial stress distribution (MPa);

σ_3 -third-order order coefficient for polynomial stress distribution (MPa);

σ_4 -fourth order coefficient for polynomial stress distribution.

To evaluate the Mode I stress intensity factor, K_I, for surface crack under thermal stresses, the procedure from ref. [18] was followed, which uses the following relation:

$$K_I\left(\frac{a}{l}\right) = \sqrt{\frac{\pi a}{Q}} \cdot \left[G_0\sigma_0 + G_1\sigma_1 \cdot \left(\frac{a}{l}\right) + G_2\sigma_2 \cdot \left(\frac{a}{l}\right)^2 + G_3\sigma_3 \cdot \left(\frac{a}{l}\right)^3 + G_4\sigma_4 \cdot \left(\frac{a}{l}\right)^4 \right], \qquad (4.63)$$

where G_0, G_1, G_2, G_3, G_4 are the influence coefficients (or magnification factors). In the case of a long axial crack and also fully circumferential crack on the inner pipe surface, the Q parameter is considered as $Q=1$.

Usually, the influence coefficient values are provided in published tables as function of the component and crack geometry, also with certain geometric/dimensional limits. In ref. [15] these limits are

$$0.0 \leq \frac{a}{l} \leq 0.8, \tag{4.64}$$

$$2 \leq \frac{r_i}{l} \leq 1000, \tag{4.65}$$

Where: a- is the crack depth, l- is the wall thickness, r_i= is the inner pipe radius. For the pipe geometry of the Civaux 1 case, the ratio in Equation (4.3.5) is:

$$\frac{r_i}{l} \approx 13. \tag{4.66}$$

Therefore, this requires a cubic spline interpolating method to be applied on labeled data in order to provide the adequate influence coefficients G_0, G_1, G_2, G_3, G_4, for Civaux 1 case geometry; the corresponding values are not found directly from Table C9 of API 579 procedure [15]. Table E1 form Appendix E displays the published values in the range of interest [15].

Based on the Table E2 (Appendix E) each influence coefficient needs a new interpolation as a function of the $\dfrac{a}{l}$ ratio, in order to consider the dependence of the stress intensity factor K_I on crack depth a . The following relationships give the required dependence:

$$G_0\left(\frac{a}{l}\right) = 10.6083 \cdot \left(\frac{a}{l}\right)^3 - 1.9273 \cdot \left(\frac{a}{l}\right)^2 + 1.2123 \cdot \left(\frac{a}{l}\right) + 1.1143, \tag{4.67}$$

$$G_1\left(\frac{a}{l}\right) = 3.5302 \cdot \left(\frac{a}{l}\right)^3 - 0.4091 \cdot \left(\frac{a}{l}\right)^2 + 0.4166 \cdot \left(\frac{a}{l}\right) + 0.6799, \tag{4.68}$$

$$G_2\left(\frac{a}{l}\right) = 1.775 \cdot \left(\frac{a}{l}\right)^3 - 0.11 \cdot \left(\frac{a}{l}\right)^2 + 0.215 \cdot \left(\frac{a}{l}\right) + 0.5234, \tag{4.69}$$

$$G_3\left(\frac{a}{l}\right) = 1.0823 \cdot \left(\frac{a}{l}\right)^3 - 0.0284 \cdot \left(\frac{a}{l}\right)^2 + 0.1364 \cdot \left(\frac{a}{l}\right) + 0.4397, \tag{4.70}$$

$$G_4\left(\frac{a}{l}\right) = 0.7044 \cdot \left(\frac{a}{l}\right)^3 + 0.056 \cdot \left(\frac{a}{l}\right)^2 + 0.0914 \cdot \left(\frac{a}{l}\right) + 0.3785. \tag{4.71}$$

A check was performed to compare the stress intensity factor calculated by the methodology described above with that from finite-element analysis results [10].

4.4.1 The SIF Frequency Response Function

The calculation of the SIF from the surface temperature variation can be regarded as a frequency response calculation with modulus $\left| H_K\left(\dfrac{a}{l},\omega\right) \right|$. The methodology was given elsewhere [16, 17], most of them using the weight function method.

The magnitude of frequency transfer function for SIF may be written in terms of the stress frequency response function assumed in Equation (4.60) [4]. For this, function $\left| H_{\sigma_\theta}(r,\omega) \right|$ is written as through-thickness profile

$$\left| H_{\sigma_\theta}\left(\frac{x}{l},\omega\right) \right| = h_0(\omega) + h_1(\omega)\cdot\left(\frac{x}{l}\right) + h_2(\omega)\cdot\left(\frac{x}{l}\right)^2 + h_3(\omega)\cdot\left(\frac{x}{l}\right)^3 + h_4(\omega)\cdot\left(\frac{x}{l}\right)^4. \quad (4.72)$$

As one can see in Figure (4.7), the fitting coefficients h_j $(j=0,...,4)$ depend on the loading frequency ω $(\omega=2\pi f)$. For the range of frequency 0.1 to 1.0 Hz, their polynomial dependences on frequency are given in Appendix E. For next convenience, the dependences are directly given as frequency in *Hz*.

The magnitude of SIF frequency response function (or modulus of the frequency transfer function for SIF) is assumed to be given by

$$\left| H_K\left(\frac{a}{l},\omega\right) \right| = \sqrt{\pi a}\cdot\left| G_K\left(\frac{a}{l},\omega\right) \right|. \quad (4.73)$$

with

$$\left| G_K\left(\frac{a}{l},\omega\right) \right| = abs\left\{ h_0(\omega)\cdot G_0\left(\frac{a}{l}\right) + h_1(\omega)\cdot G_1\left(\frac{a}{l}\right)\cdot\left(\frac{a}{l}\right) + h_2(\omega)\cdot G_2\left(\frac{a}{l}\right)\cdot\left(\frac{a}{l}\right)^2 + \right. \quad (4.74)$$

$$\left. + h_3(\omega)\cdot G_3\left(\frac{a}{l}\right)\cdot\left(\frac{a}{l}\right)^3 + h_4(\omega)\cdot G_4\left(\frac{a}{l}\right)\cdot\left(\frac{a}{l}\right)^4 \right\}$$

The relationships for $G_j\left(\dfrac{a}{l}\right)$, $j=0,...,4$ are given by Equations (4.67) to (4.71). Figure 4.8 shows the dependence of magnitude of SIF frequency response function, $\left| H_K\left(\dfrac{a}{l},\omega\right) \right|$, on loading frequency for various crack depths.

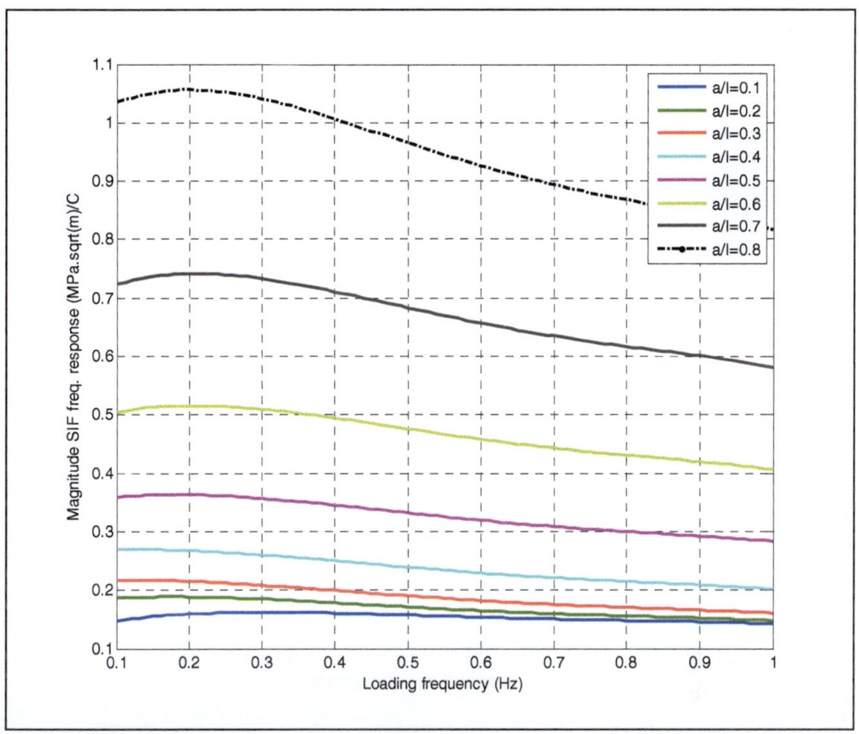

Fig. 4.8 Magnitude of SIF frequency response as function on loading frequency (*Hz*) on crack depth

As crack is growing into the thickness, the magnitude response is higher. Note that for small crack depth, the magnitude of SIF response is almost the same for the whole of frequency range, and for deeper cracks, the maximum of response is reached for 0.2-0.3 Hz.

In the next, if we convert the frequency response function into stress intensity factor, K_I, as

$$K\left(\frac{a}{l},\omega,t\right) = \Theta_0 \cdot \sqrt{\pi a} \cdot \left| G_K\left(\frac{a}{l},\omega\right) \right| \cdot \sin\left(\omega t - \varphi\right), \qquad (4.75)$$

we are able to find its dependence on loading frequency, also for various crack depths (Figure 4.9). The examination of this behavior of K_I, which is calculated for instant of time $t=T/4$ (with $T=$ time period of loading), suggests the highest its value for frequency $f=0.3$ *Hz*, which is in a good agreement with previous study [11].

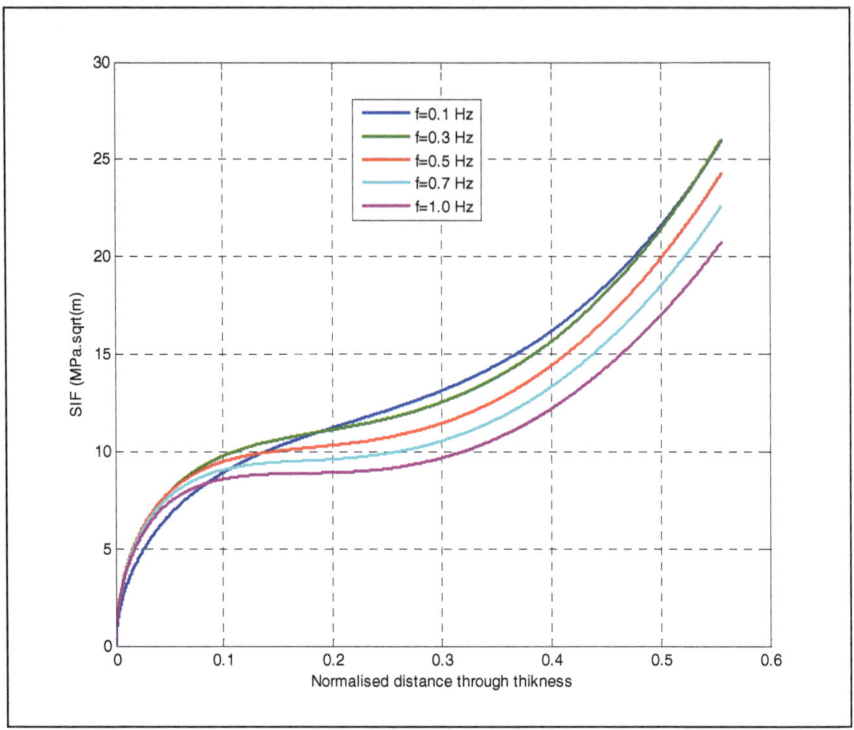

Fig. 4.9 Stress intensity factor (instant T/4) using SIF frequency response function versus crack depth

For the sake of simplicity, we designate the crack depth to the thickness ratio as:

$$x_a = \frac{a}{l}, \tag{4.76}$$

where the new normalized coordinate x_a originates at the inner pipe surface, and reminds that a is crack depth of long shallow axial crack; l is the thickness of the wall. By using the normalized coordinate, x_a , then the modulus of the frequency transfer function for SIF becomes

$$\left|H_K\left(x_a,\omega\right)\right| = \sqrt{\pi \cdot l \cdot x_a} \times abs\left\{\sum_{j=1}^{4} h_j\left(\omega\right) \cdot G_j\left(x_a\right) \cdot \left(x_a\right)^j\right\}. \tag{4.77}$$

The Equation (4.3.17) will be used further in application even it is based on a single sinusoidal temperature loading.

4.4.2 Power Spectral Density of SIF and Its Spectral Moments

The analysis above has been performed by considering a sinusoidal thermal loading of surface temperature fluctuations. For mixing tees the surface temperature variation will be random. For the present analysis, the input of surface temperature fluctuations is characterized by its power spectral density (PSD), which is the Fourier transform of the autocorrelation function. This may be obtained from experimental measurements.

It is necessary to postulate a probability distribution functional for temperature. This will be taken to be Gaussian, implying a Gaussian probability density function for temperature at any instant [17]. As it was shown in previous chapters, a Gaussian stochastic process is completely described by its PSD.

The approach followed here is to consider the temperature fluctuation and its spectrum as a Gaussian stationary narrow-band process. In this case, following the Equation (3.101), the magnitude of SIF frequency response function, $\left| H_K \left(x_a, \omega \right) \right|$, relates the PSD of SIF, $S_K \left(x_a, \omega \right)$, and PSD of surface temperature $S_T(\omega)$, respectively, as

$$S_K \left(x_a, \omega \right) = \left| H_K \left(x_a, \omega \right) \right|^2 S_T \left(\omega \right). \tag{4.78}$$

The mean square (variance) of the SIF is given by

$$K_{rms}^2 \left(x_a \right) = \int_{-\infty}^{\infty} S_K \left(x_a, \omega \right) d\omega. \tag{4.79}$$

If it is assumed that temperature's PSD is a stationary Gaussian narrow band process, then follows that this PSD has relevant values into a limited frequency interval. Moreover, from practical point of view here is considered the one-sided PSD with frequency expressed in Hertz (cycles/second)

$$W_T \left(f \right) = 4\pi \cdot S_T \left(\omega \right), \tag{4.80}$$

with $W_T(f)$ expressed in $(^\circ C)^2/Hz$, and $f = \dfrac{\omega}{2\pi}$ in Hz.

For the sake of simplicity, we shall consider the model of the PSD from Figure 3.4 (a), in a more idealized shape as it is displayed in Figure 4.10. This yields to a mean square of temperature signal (variance) in the form

$$R_T \left(\tau = 0 \right) = \sigma_T^2 = \int_{f_1}^{f_2} W_T \left(f \right) df = W_{0T} \left(f_2 - f_1 \right) = W_{0T} \Delta f, \tag{4.81}$$

for a normalized stationary Gaussian random process.

By considering the one-sided PSD with frequency f expressed in Hertz then the PSD of SIF is given by

$$W_K \left(x_a, f \right) = \left| H_K \left(x_a, f \right) \right|^2 W_T \left(f \right). \tag{4.82}$$

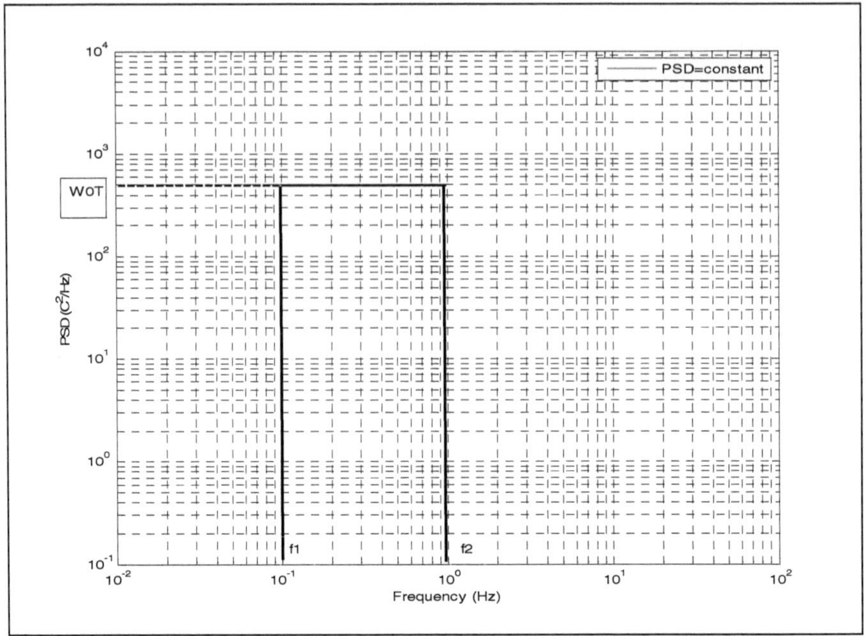

Fig. 4.10 One-sided PSD for temperature fluctuations

The PSD of the thermal spectrum from Figure 4.10, can be written as

$$W_T\left(f\right)=\begin{cases}W_{0T}=ct, f\in\left[f_1,f_2\right]\\ 0, f\in\left[0,f_1\right)\cup\left(f_2,\infty\right]\end{cases}.\tag{4.83}$$

This in turn that one-sided PSD of K_I becomes

$$W_K\left(x_a,f\right)=\left|H_K\left(x_a,f\right)\right|^2 W_{0T}.\tag{4.84}$$

The Equation (4.84) will be used in the next application, but one should note that a more general form for PSD of SIF can be manipulated in the same way to obtain the expectation of crack propagation rate under a specific thermal spectrum.

The zero-order moment of K_I PSD, which means variance of SIF for a normalized stationary Gaussian stochastic process, is given by

$$K_{rms}^2\left(x_a,\tau=0\right)=\int_{f_1}^{f_2}W_K\left(x_a,f\right)=\int_{f_1}^{f_2}\left|H_K\left(x_a,f\right)\right|^2 W_{0T}df=W_{0T}\int_{f_1}^{f_2}\left|H_K\left(x_a,f\right)\right|^2 df.\tag{4.85}$$

The stochastic properties for a stationary Gaussian stochastic process are also valid for SIF fluctuations. Thus, because the Gaussian stationary narrow-band process is smooth and harmonic, for every peak, there is a corresponding zero up-crossing (see Equation 3.70), and the expected rate of zero up-crossing rates of K_I is

$$\dot{N}_{K,0} = \dot{N}_{K,p} = \frac{\sqrt{\int_{f_1}^{f_2} f^2 W_K(x_a, f)\, df}}{\sqrt{\int_{f_1}^{f_2} W_K(x_a, f)\, df}} = \frac{\sqrt{\int_{f_1}^{f_2} f^2 \left| H_K(x_a, f) \right|^2 df}}{\sqrt{\int_{f_1}^{f_2} \left| H_K(x_a, f) \right|^2 df}}. \tag{4.86}$$

The frequency of peaks of any magnitude for K_I, which is supposed to be a stationary narrow-band Gaussian process, is characterized by Rayleigh's distribution, as follows from Equation 3.80:

$$f_{K_I}(x_a, \tau = 0) = \frac{K_I}{K_{rms}^2(x_a, \tau = 0)} \exp\left[-\frac{1}{2} \frac{K_I}{K_{rms}^2(x_a, \tau = 0)} \right]. \tag{4.87}$$

Equation (4.87) can be written in a complete form as function of magnitude of SIF frequency response, by taking into account the Equation (4.85):

$$f_{K_I}(x_a, \tau = 0) = \frac{K_I}{W_{0T} \int_{f_1}^{f_2} \left| H_K(x_a, f) \right|^2 df} \exp\left[-\frac{1}{2} \frac{K_I}{W_{0T} \int_{f_1}^{f_2} \left| H_K(x_a, f) \right|^2 df} \right]. \tag{4.88}$$

Based on the knowledge of SIF frequency response magnitude, the Equation (4.88) allows to obtain the mth moment of Rayleigh's distribution.

4.5   Lifetime Estimation for Thermal Fatigue Crack Growth in Mixing Tees

Present analysis assumes a Paris law for crack growth per cycle

$$\frac{da}{dN} = C \cdot (\Delta K)^n, \tag{4.89}$$

where N is the number of maxima and ΔK is the range between the maximum and next minimum; here we consider the range between maximum and next zero

$$\Delta K = K. \tag{4.90}$$

The stochastic model for thermal fatigue crack growth developed to here include a first part for incorporating stochastic loads (derived into stochastic behavior of K) and a second one, that deals with Monte Carlo simulation to accommodate statistical characteristics of crack growth under constant amplitude (C and n Paris law parameters).

4.5.1 Expected Value of Crack Growth Rate

Time-dependent fluctuation of temperature should be correlated with time dependent fluctuation of crack growth from Paris's law. Because the number of loading cycles is a discrete variable with respect to the time variable, the number of loading cycles is modified into a continuous variable by introducing an average cyclic rate. So, when time-dependent stochastic analysis is conducted, the crack growth rate of a random flaw size, a, should be written in the following form:

$$\frac{da}{dt} = \frac{da}{dN}\frac{dN}{dt} = v_p\frac{da}{dN}, \tag{4.91}$$

where v_p is the mean rate of maxima, that is constant for a Gaussian stationary stochastic process. Moreover, for this kind of process v_p may be identified with expected rate of up-crossing rate (equal to expected rate of peak crossing)

$$v_p = \dot{N}_{K,0} = \dot{N}_{K,p} = \frac{\sqrt{\int_{f_1}^{f_2} f^2 W_K(x_a,f)\,df}}{\sqrt{\int_{f_1}^{f_2} W_K(x_a,f)\,df}} = \frac{\sqrt{\int_{f_1}^{f_2} f^2 |H_K(x_a,f)|^2\,df}}{\sqrt{\int_{f_1}^{f_2} |H_K(x_a,f)|^2\,df}}, \tag{4.92}$$

with the significance of parameters mentioned in previous chapter.

We assume a linear summation of damage and ignore the effect of positive minima [17], and in this case the expectation rate of crack growth in respect to cycles is

$$E\left[\frac{da}{dN}\right] = \int_0^\infty C\cdot(K)^n f_{K_I}(x_a,\tau=0)\,dK. \tag{4.93}$$

By inserting Equation (4.86) and re-arranging the Equation (4.93), results in

$$E\left[\frac{da}{dN}\right] = C\cdot\int_0^\infty (K)^n \frac{K_I}{K_{rms}^2(x_a,\tau=0)}\exp\left[-\frac{1}{2}\frac{K_I}{K_{rms}^2(x_a,\tau=0)}\right]dK. \tag{4.94}$$

This last form is the nth moment of Rayleigh's distribution, concerned on K_I:

$$E\left[\frac{da}{dN}\right] = C\cdot\left[K_{rms}(x_a,\tau=0)\right]^n\cdot 2^{\frac{n}{2}}\Gamma\left(1+\frac{1}{n}\right), \tag{4.95}$$

where Γ is the Gamma function. Using the magnitude of SIF frequency response yields to

$$E\left[\frac{da}{dN}\right] = C \cdot \left[W_{0T} \int_{f_1}^{f_2} \left|H_K\left(x_a, f\right)\right|^2 df\right]^n \cdot 2^{\frac{n}{2}} \Gamma\left(1 + \frac{1}{n}\right). \tag{4.96}$$

By inserting Equations 4.4.4 and 4.4.8 into Equation 4.4.3, the final form of stochastic crack growth rate is

$$\frac{dx_a}{dt} = \frac{C}{l} \cdot \frac{\sqrt{\int_{f_1}^{f_2} f^2 \left|H_K\left(x_a, f\right)\right|^2 df}}{\sqrt{\int_{f_1}^{f_2} \left|H_K\left(x_a, f\right)\right|^2 df}} \left[W_{0T} \int_{f_1}^{f_2} \left|H_K\left(x_a, f\right)\right|^2 df\right]^n \cdot 2^{\frac{n}{2}} \Gamma\left(1 + \frac{1}{n}\right), \tag{4.97}$$

with l being the wall thickness.

This equation may be integrated numerically to obtain the crack length, or x_a, as a function of time, when C and n a given deterministically.

4.5.2 Coupling of the Stochastic Model with the Probabilistic Input to Assess Crack Growth Lifetime

A prospective study for the probabilistic approach of thermal fatigue in mixing tees (Civaux 1 damage case) by limit state function and Monte Carlo simulation, based on sinusoidal approach, has been done in a previous work [18]. In the present work, the limit state function will be based on the Equation (4.97), and in the next, we describe shortly the methodology used for probabilistic input to account the variability in initial crack depth and in C scaling parameter so far.

A limit state is generally defined as a state of a structure or part of the structure that no longer meets the requirements laid down for its performance or operation. In another way, the limit states can be defined as a specific set of states that separate a desired state from an undesirable state which fails to meet the design requirements. In addition, more generally, we can say that a *Limit State* is a mathematical criterion that categorizes any set of values of the relevant structural variables (loads, material and geometrical variables) into one of two categories: a "safe" set and a "failure" set.

A limit state of thermal fatigue damage due to thermal loading is considered a crack penetration depth of 80% wall thickness. It is possible to define the failure function or limit state function as a function of the number of thermal fatigue cycles, N_{cycles}, as [18]

$$g\left(N_{cycles}\right) = a_{cr} - a_f\left(N_{cycles}\right), \tag{4.98}$$

or equivalent

$$g(N_{cycles}) = 1 - \frac{a_f(N_{cycles})}{a_{cr}},\tag{4.99}$$

where a_{cr} is a critical depth of the fatigue crack, corresponding to 80% of the wall-thickness, and $a_f(N_{cycles})$ is the final crack depth after N cycles of thermal loading. The failure is predicted when the number of cycles N_{cycles}, will produce the following condition

$$g(N_{cycles}) \leq 0,\tag{4.100}$$

which means $a_f > a_{cr.}$, failure condition.

To combine the stochastic behavior of K with statistical characteristics of crack growth under constant amplitude (C and n Paris law parameters), and also with initial crack depth distribution, we define the limit state function in the form

$$g(t_{ref}) = 1 - \frac{t_{ref}}{t_{stoch}},\tag{4.101}$$

where: t_{ref} is the reference time period for thermal fatigue crack growth under thermal spectrum; tstoch is estimated values of the lifetime for stochastic crack growth derived from Equation (4.97). Thus, for integrating, Equation (4.97) can be written as

$$t_{stoch} = \int_{x_{ai}}^{0.8} \left[\frac{C}{l} \cdot \frac{\left(\int_{f_1}^{f_2} f^2 |H_K(x_a,f)|^2 \, df \right)^{\frac{1}{2}}}{\left(\int_{f_1}^{f_2} |H_K(x_a,f)|^2 \, df \right)^{\frac{1}{2}}} \cdot \left[W_{0T} \int_{f_1}^{f_2} |H_K(x_a,f)|^2 \, df \right]^n \cdot 2^{\frac{n}{2}} \Gamma\left(1+\frac{1}{n}\right) \right]^{-1} dx_a,\tag{4.102}$$

where x_{ai} is a sample from initial crack distribution, which is supposed to be known; the final crack size uses the same failure criterion as above, means $x_{acr} = a_{cr}/l = 0.8$. Note that in the present study, we consider long cracks; as a consequence, the limit state function is referred just to the crack depth.

During the Monte Carlo simulation (MCS), the trials which satisfy condition

$$g(t_{ref}) \leq 0,\tag{4.103}$$

are accounted as n_{fail} and the probability of failure for a certain period of time (t_{ref}) is given by

$$P_f = \frac{n_{fail}}{N_{trials}},\tag{4.104}$$

where N_{trials} is the total number of trials' simulation.

References

[1] Yung-Li, L., et al.: Fatigue testing and analysis (Theory and Practice). Elsevier Butterworth–Heinemann (2005)

[2] Yao, J.T.P., Kozin, F., et al.: Stochastic fatigue, fracture and damage analysis. Structural Safety 3, 231–267 (1986)

[3] Jones, I.S.: The frequency response model of thermal striping for cylindrical geometries. Fatigue and Fracture of Engineering Materials and Structures 20(6), 871–882 (1997)

[4] Jones, I.S., Lewis, M.W.: A frequency response method for calculating stress intensity factors due to thermal striping loads. Fatigue and Fracture of Engineering Materials and Structures 17(6), 709–720 (1994)

[5] Miller, A.G.: Crack propagation due to random thermal fluctuation: effect of temporal incoherence. International Journal of Pressure vessels and Piping 8, 15–24 (1980)

[6] Miller, A.G.: Equivalent strain range due to random thermal fluctuations: effect of spatial incoherence. International Journal of Pressure vessels and Piping 8, 105–130 (1980)

[7] Radu, V., Taylor, N., Paffumi, E.: Development of new analytical solutions for elastic thermal stress components in a hollow cylinder under sinusoidal transient thermal loading. International Journal of Pressure Vessels and Piping 85, 885–893 (2008)

[8] Radu, V., Paffumi, E., Taylor, N.: New analytical stress formulae for arbitrary time dependent thermal loads in pipes. European Commission Report EUR 22802 DG JRC, Petten, NL (June 2007)

[9] Paffumi, E., Radu, V.: Status on the knowledge on cracks evolution under loadings from a thermal spectrum-Crack propagation and possible arrest/penetration, published as European Commission Report, European Communities (2009)

[10] Radu, V., Paffumi, E., Taylor, N., Nilsson, K.-F.: Assessment of thermal fatigue crack growth in the high cycle domain under sinusoidal thermal loading – An application-Civaux 1 case, published as European Commission Report EUR 23223 EN, DG JRC, Petten, The Netherlands (2007), doi:DOI: 10.2790/4943 ISSN: 1018-5593, ISBN: 978- 92-79-08218-4

[11] Radu, V., Paffumi, E., Taylor, N., Nilsson, K.-F.: A study on fatigue crack growth in high cycle domain assuming sinusoidal thermal loading. International Journal of Pressure Vessels and Piping (2009), doi:10.1016/j.ijpvp, 10.007

[12] Paffumi, E., Radu, V.: Status on the knowledge on cracks evolution under loadings from a thermal spectrum-Crack

[13] Propagation and possible arrest/penetration, published as European Commission Report, European Communities (2009)

[14] Noda, N., Hetnarski, R.B., Tanigawa, Y.: Thermal Stresses, 2nd edn. Taylor & Francis (2003)

[15] Heller, R.A., Thangjitham, S.: Probabilistic Methods in Thermal Stress Analysis. In: Hetnarski, R.B. (ed.) Thermal Stress II, Elsevier Science Publishers B.V (1987)

[16] API 579, Fitness-for-Service-API Recommended Practice 579, 1st edn. American Petroleum Institute (January 2000)

[17] Jones, I.S., Lewis, M.W.: A frequency response method for calculating stress intensity factors due to thermal striping loads. Fatigue and Fracture of Engineering Materials and Structures 17(6), 709–720 (1994)

[18] Miller, A.G.: Crack propagation due to random thermal fluctuation: effect of temporal incoherence. International Journal of Pressure vessels and Piping 8, 15–24 (1980)

[19] Radu, V., Paffumi, E., Taylor, N., Nilsson, K.-F.: A prospective study for probabilistic approach of thermal fatigue in mixing tees, published as European Commission Report EUR 23570EN, ISSN: 1018-5593, DG JRC, Petten, The Netherlands (2009)

Chapter 5
Application

Abstract. The thermal fatigue damaging case from 1998, when a longitudinal crack was discovered at the outer edge of an elbow in a mixing zone of the Residual Heat Removal System (RHRS) of the Civaux NPP unit 1 has been chosen to verify the stochastic model from the book. The variability in statistical properties of material parameters is usually accounted by the statistical properties of Paris law parameters C and n. Also, the original crack depth of flaws has a certain probability density function, which is more related to probability of detection based on experimental in-service inspection (ISI) results. A lognormal probability density function of C scaling parameter and an exponential one for initial crack depth are used to provide a probabilistic input for solving the integral giving the crack depth as a function of time. The results of the stochastic approach to modeling of thermal fatigue crack growth in mixing tee, completed with probabilistic input to account variability in material characteristics, are given as the probability of failure as a function of the time reference period.

5.1 Application of the Stochastic Model to Assess the Thermal Fatigue Crack Growth

The previous methodology to evaluate stochastic crack growth under thermal fluctuation will be applied to Civaux1 case, which is a very well known damaging case on the mixing tees concern. It is shortly described in the Appendix B. The limit state function defined by Equation (4.101) requires the probability density functions for initial crack depth and also for C and n parameters from Paris's law.

The initial crack size distribution has a very strong influence on the deterministic and also probabilistic assessment of a component lifetime. Usually, the initial crack distribution involves three kinds of distributions: - crack depth distribution, crack aspect ratio distribution, crack existence frequencies.

The present approach considers only cracks that start out as long inner surface cracks, characterized by initial crack depth distribution as an exponential distribution. The corresponding probability density function is

$$f(z;\mu) = \frac{1}{\mu} e^{-\frac{x}{\mu}}, \, z{\geq}0; \, \mu{>}0, \tag{5.1}$$

© Springer International Publishing Switzerland 2015

V. Radu, *Stochastic Modeling of Thermal Fatigue Crack Growth*,

Applied Condition Monitoring 1, DOI: 10.1007/978-3-319-12877-1_5

In the present book, we adopted the pdf from equation (5.1) in the following form

$$p\left(a;\mu_a\right)=\frac{1}{\mu_a}e^{-\frac{a}{\mu_a}},\ 0\le a\le a_0.\tag{5.2}$$

For a pipe thickness l=9 mm as in the present work, we consider a mean value of the crack depth as μ_a=1 mm and the coefficient of variation in this case is CoV=1. The mean value of the crack depth is small, but in deterministic assessments, this value is generally assumed as a started depth for fatigue crack growth. The proposed value to be used for a_0 in Equation (5.2), which usually is considered ∞ in the case of the thick pipe-wall, can be seen as cracks detected by ISI, before assessment. Therefore, in this application, a value of three mm is chosen, which means about 30% of the wall pipe. In this way, the cracks with depth bigger than three mm, generated by MC sampling from exponential distribution, are not accounted for crack growing.

The fatigue crack growth rates are calculated in the deterministic assessments using stainless steel crack growth law given in ASME:

$$\frac{da}{dN}=C\cdot\left(\Delta K_I\right)^n,\tag{5.3}$$

where n=3.3 is the slope of the $log(da/dN)$ versus $log(\Delta K_I)$ and C is a scaling parameter. Slopes (n) and intercepts (C) for all fatigue data are usually highly correlated. Ignoring this correlation can give misleading results in a simulation. An alternate method to account for this correlation is to use a constant slope and put all of variability into the intercept [1]. For a constant slope, the variability in fatigue lives will be directly related to variability in the material constant C. The scatter in the experimental fatigue data is represented by a lognormal distribution for C scaling parameter, with the following pdf

$$f\left(z;\mu_0,\sigma_0\right)=\frac{1}{z\cdot\sigma_0\sqrt{2\pi}}e^{-\frac{1}{2}\left(\frac{\ln z-\mu_0}{\sigma_0}\right)^2},\tag{5.4}$$

where μ_0, σ_0, are parameters connected with the median/mean value and standard deviation as shown in Appendix F.

From reference [1] we consider the following parameters for lognormal distribution of C:

- Median: C_{median}=10.04$\cdot$10^{-12} (m/cycle/ MPa$\sqrt{}$m),   (5.5)
- Standard deviation: σ_C=2.2$\cdot$10^{-11}.   (5.6)

With these parameters, a mean value is derived as C_{mean}= 1.664$\cdot$10^{-11} (m/cycle/ MPa$\sqrt{}$m) and coefficient of variation by CoV=1.32. The log-normal distribution is used to produce random values for C constant in Paris's law with the following sequence in MATLAB Statistics Toolbox:

$$C = F^{-1}\left(\Phi(u);\mu_0,\sigma_0\right);\ \mu_0 = -25.3244;\ \sigma_0 = 1.0053. \tag{5.7}$$

Here $F^{-1}\left(\Phi(u);\mu_0,\sigma_0\right)$ is the MATLAB function for inverse function of log-normal CDF. The parameters μ_0 and σ_0 were derived with relationships from Appendix F.

The stochastic approach presented in previous chapters requires knowledge of statistical properties of the thermal spectrum. The Civaux case was characterized by a temperature fluctuation [2], which is displayed in Figure 5.1.

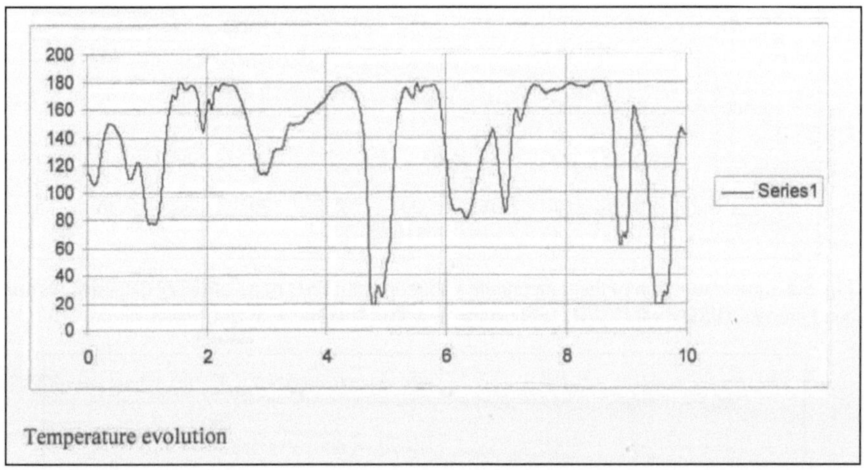

Fig. 5.1 Temperature fluctuations in a pipe where a thermal fatigue crack penetrated the wall

For a thermal spectrum assumed to be a stationary Gaussian stochastic process, we use the one-sided temperature PSD, see figure 4.3.3. In this case, we choose a value for its PSD

$$W_T(f) = \begin{cases} W_{0T} = 500C^2\,/\,Hz, f \in [0.1,1.0] & Hz \\ 0, f \in [0,0.1Hz) \cup (1.0Hz,\infty] \end{cases}. \tag{5.8}$$

By re-conversion using RSA method (Random Spectral Amplitudes) described in chapter 3.2.1, we extract a sample function for temperature that is displayed in Figure 5.2 with zero-mean, and in Figure 5.3 with a non-zero mean. For a shorter interval, 10 seconds, Figure 5.4 shows a quite similar aspect to those reflected in Figure 5.1.

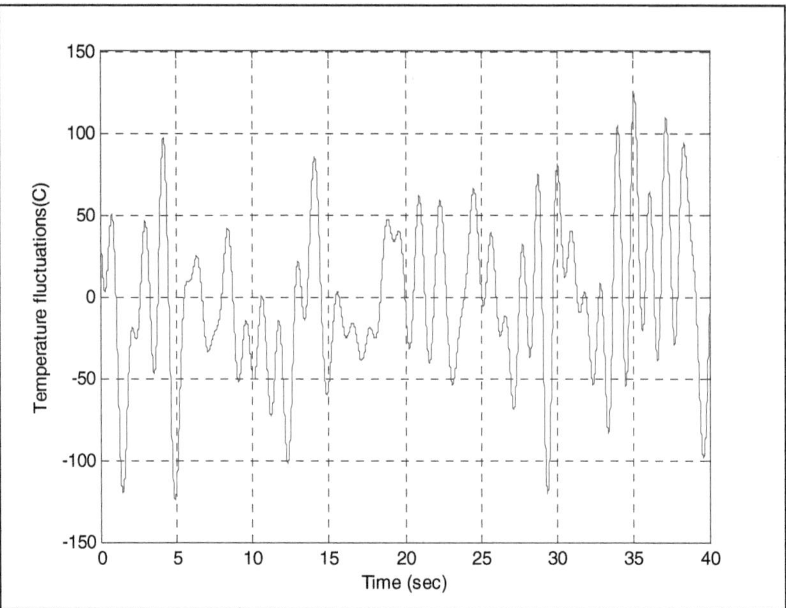

Fig. 5.2 Sample function of the temperature variation from PSD of a stationary Gaussian narrow-band process, (PSD=500 C^2/Hz, frequency range=0.1-1.0 Hz, with zero-mean value)

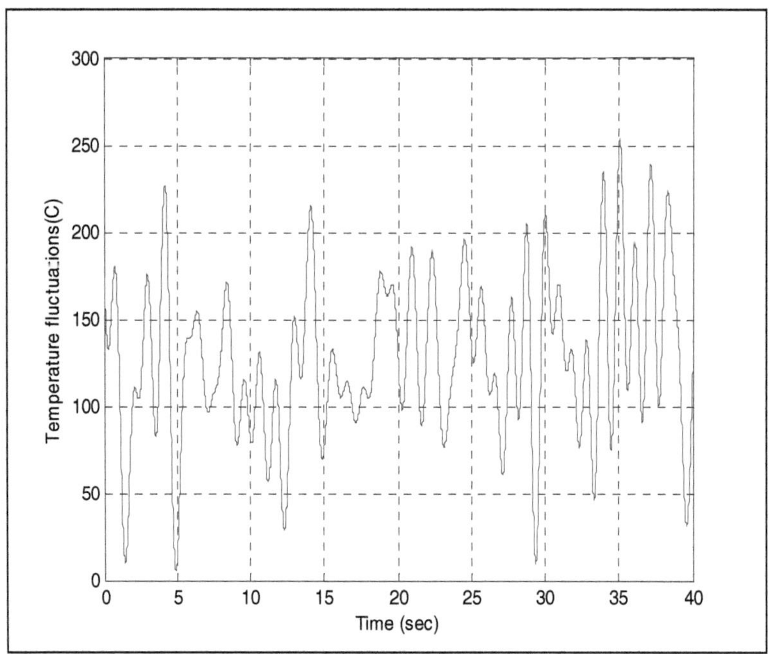

Fig. 5.3 Sample function of the temperature variation from PSD of a stationary Gaussian narrow-band process, (PSD=500 C^2/Hz, frequency range=0.1-1.0 Hz, with non-zero mean value)

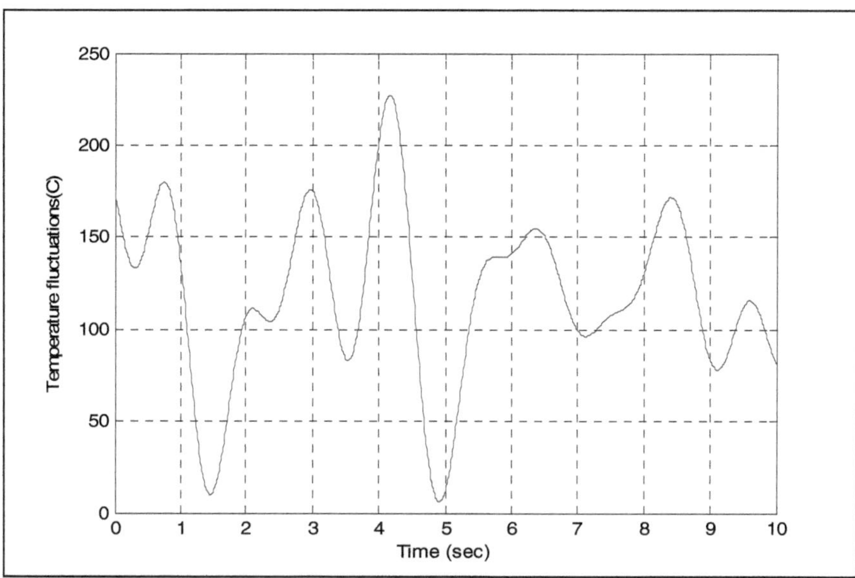

Fig. 5.4 A shorter time interval sequence of temperature fluctuations (from Figure 5.3)

The Monte Carlo analyses were performed by implementing in the MATLAB environment specific scripts and described function, using a number of trials in the range of 10^4-10^5. Typical histograms for both the initial crack depth and C scaling parameter are illustrated in Figure 5.5 and Figure 5.6, respectively.

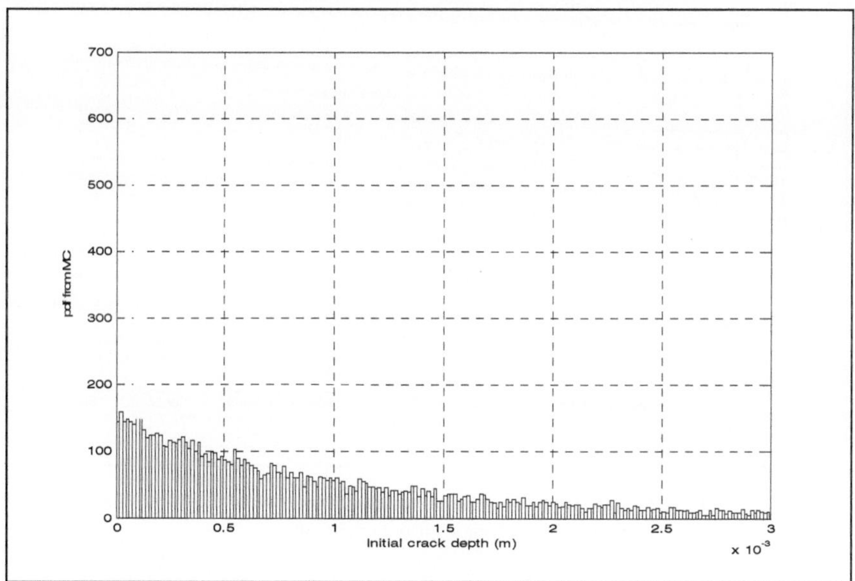

Fig. 5.5 Probability density function for initial crack depth (MC simulations)

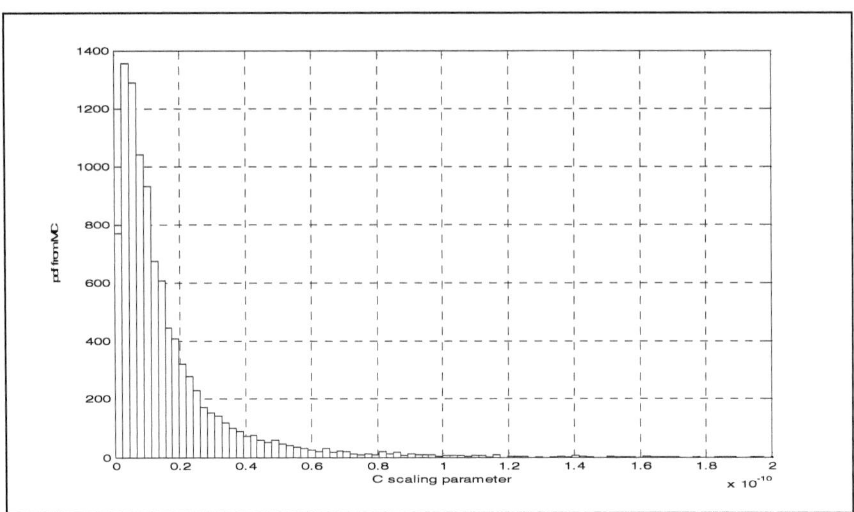

Fig. 5.6 Probability density function for *C* scaling parameter (MC simulations)

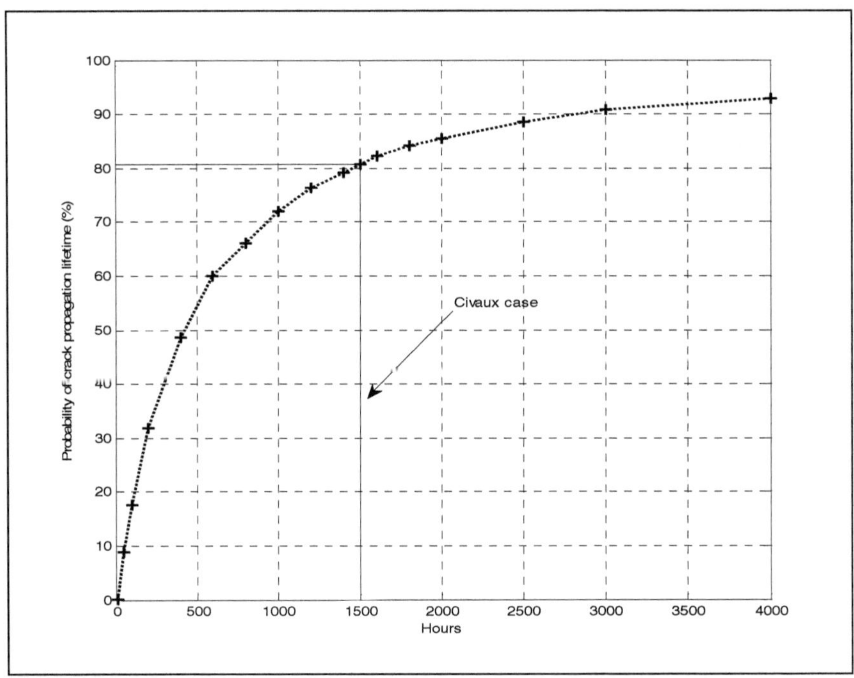

Fig. 5.7 Probabilities of failure: the stochastic modeling results of fatigue crack growth coupled with probabilistic input for Monte Carlo simulation

The results from analyses of stochastic modeling of fatigue crack growth due to thermal spectrum and using probabilistic input to account variability in Paris's law parameters and initial crack depth are displayed in Figure 5.7. The probabilities of failure, actually defined mathematical by limit state function satisfying Equation (4.101) are given as a function of the reference time period. It is shown also, on the same graph, the lifetime for crack penetration through wall-piping in Civaux case. One can see that the time of 1500 hours, which has been referred as time of the crack penetration through the wall in an elbow region of Civaux pipe, corresponds to a probability of failure about 80%. This value of the probability is quite high, and we can conclude that stochastic modeling of thermal fatigue cracks growth, coupled with probabilistic input, gives reasonable assessment of the event, even it is performed in the elastic domain.

References

[1] Gosselin, S.R., Simonen, F.A., Heasler, P.G., Doctor, S.R.: Fatigue Crack Flaw Tolerance in Nuclear Power Plant Piping; A basis for Improvements to ASME Code Section XI Appendix L, NUREG/CR-6934, PNNL-16192 (May 2007)
[2] Gourdin, C., Marie, S., Chapuliot, S.: An Analytical Thermal Fatigue crack growth approach, SMiRT 20-Division 2, Paper 1796, Espoo, Finland, August 9-14 (2009)

Chapter 6
Conclusions

- The book proposes a stochastic model to assess thermal fatigue crack growth in mixing tees of NPP with the temperature spectrum assumed to be a Gaussian stationary narrow-band stochastic process. The stochastic crack growth model includes a main part for incorporating randomness in service loads, and also another one, which includes a description of statistical characteristics of crack growth under constant amplitude loadings.

- Based on the analytical solution of temperature response (Hankel transform) within SIN-methodology, developed in past work a temperature frequency response function through pipe thickness is developed. By considering the analytical solution for thermal stresses also developed in preceding work, a stress frequency response for thermal hoop stress is derived and based on this an SIF frequency response magnitude is obtained. For a one-sided PSD model of temperature fluctuation, the PSD of SIF is obtained, by FRF methodology. Consequently, the expected value of crack growth rate in HCF domain can be assessed using the Rayleigh's distribution moments. The variability in Paris's law parameter is accounted and in the crack distribution as well, and the probabilities of failure are obtained by MCS. The methodology is used to evaluate stochastic crack growth under thermal fluctuation in the NPP Civaux1 case, which is a very well known damaging case on the mixing tees concern.

- The present methodology based on the stochastic modeling of thermal fatigue crack growth can be used to analyze and improve the screening criteria proposed to avoid cracking issues in nuclear piping, especially in tee connection where turbulent mixing of flows with different temperature occur.

© Springer International Publishing Switzerland 2015

V. Radu, *Stochastic Modeling of Thermal Fatigue Crack Growth*,

Applied Condition Monitoring 1, DOI: 10.1007/978-3-319-12877-1_6

Appendices

Appendix A
The Use of Finite Hankel Transform

The method of integral transforms may be applied to the solution of boundary-value problems in mathematical physics [1]. Let us consider a function $f(x)$ defined on a prescribed finite interval (a, b). Furthermore, we consider finite transforms of the kind

$$\mathbf{T}[f(x);\xi] = \int_a^b p(x)K(x,\xi)f(x)\,dx\,, \qquad (A1)$$

where the prescribed function $p(x)$ is of nature of a "weight function" [1] – in sense of the theory of orthogonal polynomials- and function $K(x,\xi)$ denotes a prescribed function of x on the open interval (a,b) for each value of the parameter ξ, whose domain is also prescribed ($\mathbf{T}$ is a linear operator).

For a particular problem, the form of the kernel $K(x,\xi)$ requires consideration of the solutions of the self-ad joint differential equation

$$\mathbf{L}z = \frac{1}{p(x)}\left\{\frac{d}{dx}q(x)\frac{dz}{dx}\right\} + l(x)z(x) = 0\,, \qquad (A2)$$

where $q(x)\in C^2[a,b]$ and it is assumed that $q(x)/p(x)>0$ for $x\in[a,b]$. We consider that $f\in C^1[a,b]$ and introduce the boundary operators $\mathbf{M}$ and $\mathbf{N}$ define as

$$\mathbf{M}f = a_1 f(a) + a_2 \left.\frac{\partial f}{\partial x}\right|_{x=a}\,, \qquad (A3)$$

$$\mathbf{N}f = b_1 f(a) + b_2 \left.\frac{\partial f}{\partial x}\right|_{x=b}\,, \qquad (A4)$$

where a_1, a_2, b_1 and b_2 are prescribed constants. Note that at least one of the a_1 and a_2 and at least one of b_1 and b_2 is nonzero [1]. If we denote by $K(x,\xi_n)$ and ξ_n the normalized eigenfunction and eigenvalue of Sturm-Liouville problem

$$(L-\xi)K = 0, \qquad \mathbf{M}K = \mathbf{N}K = 0\,, \qquad (A5)$$

then the finite transform defined by equation

$$T\left[f;\xi_n\right] \equiv \overline{f}_n = \int_a^b p(x)K(x,\xi_n)dx,$$ (A6)

has the property that if $f \in C^l[a,b]$ then

$$T\left[\mathbf{L}f;\xi_n\right] = \xi_n\overline{f}_n + \frac{q(a)}{a_1}\left.\frac{\partial K(x,\xi_n)}{\partial x}\right|_{x=a}\mathbf{M}f - \frac{q(b)}{b_1}\left.\frac{\partial K(x,\xi_n)}{\partial x}\right|_{x=b}\mathbf{N}f.$$ (A7)

In deriving the equation it was considered that $a_2=0$ and $b_2=0$ [1]. The corresponding inversion formula is

$$T^{-1}\left[\overline{f}_n;x\right] \equiv f(x0 = \sum_{n=1}^{\infty}\overline{f}_nK(\xi_n,x).$$ (A8)

The finite transform corresponding to the linear operator

$$\mathbf{L} = \frac{d^2}{dr^2} + \frac{1}{r}\frac{d}{dr} - \frac{\upsilon^2}{r^2}.$$ (A9)

which is of the form of Equation (A2) with $p(x)=r$, $q(x)=r$, $l(x)=-\upsilon^2r^2$ are called finite Hankel transforms [1]. With $\xi=-s^2$, the first of Equations (A5) is Bessel's equation

$$\frac{d^2K}{dr^2} + \frac{1}{r}\frac{dK}{dr} + \left(s^2 - \frac{\upsilon^2}{r^2}\right) = 0.$$ (A10)

If we consider the case

$$a \le r \le b, \qquad \mathbf{M}f = f(a), \qquad \mathbf{N}f = f(a)$$ (A11)

then required solution of Equation (A10) is

$$K(r,s_n) = J_\upsilon(s_nr)Y_\upsilon(s_na) - J_\upsilon(s_na)Y_\upsilon(s_nr),$$ (A12)

where s_n is a root of transcendental equation

$$J_\upsilon(s_nb)Y_\upsilon(s_na) - J_\upsilon(s_na)Y_\upsilon(s_nb) = 0.$$ (A13)

Here $J_\upsilon(z)$, $Y_\upsilon(z)$, are Bessel function of first and second kind of order υ.

The finite Hankel transform of third kind is defined as in reference [1] by Equation

$$\overline{f}_n \equiv H\left[f(r);\xi_n\right] = \int_a^b rf(r)\left[J_\upsilon(s_nr)Y_\upsilon(s_na) - J_\upsilon(s_na)Y_\upsilon(s_nr)\right]dr.$$ (A14)

By using the orthogonality condition for the kernel defined by Equation (A12), [1, 2], then

$$\int_a^b r\left[K\left(r,s_n\right)\right]^2 dr = \frac{2}{\pi^2 s_n^2}\frac{\left[J_v\left(s_n a\right)\right]^2 - \left[J_v\left(s_n b\right)\right]^2}{\left[J_v\left(s_n b\right)\right]^2}. \tag{A15}$$

The solution f(r) may be obtained by the inversion theorem for the operator H, [1] as

$$f\left(r\right) \equiv H^{-1}\left[\overline{f}_n;r\right] = \frac{\pi^2}{2}\sum_{n=1}^{\infty}\frac{s_n^2 J_v^2\left(s_n b\right)\overline{f}_n}{\left[J_v\left(s_n a\right)\right]^2 - \left[J_v\left(s_n b\right)\right]^2}\left[J_v\left(s_n r\right)Y_v\left(s_n a\right) - J_v\left(s_n a\right)Y_v\left(s_n r\right)\right], \tag{A16}$$

the sum extending over all the positive roots of Equation (A13). From Equation (A7) the following relation

$$H\left[\mathbf{L}f\left(r\right);s_n\right] = \frac{2J_v\left(s_n a\right)}{\pi J_v\left(s_n b\right)}f\left(b\right) - \frac{2}{\pi}f\left(a\right) - s_n^2 \overline{f}_n, \tag{A17}$$

completes the solution for a particular problem.

References

[1] Sneddon, I.N.: The Use of Integral Transforms. McGraw-Hill, New York (1993)
[2] Cinelli, G.: An extension of the finite Hankel transform and application, Pergamon Press. International Journal of Engineering Science 3, 539–559 (1965)

Appendix B
Description of the Civaux 1 Case

The main characteristics of the piping system from Civaux 1 case had been described in [1,2]. Some of the features concerning on this thermal fatigue damaging case are given in the following. In 1998, a longitudinal crack was discovered at the outer edge of an elbow in a mixing zone of the Residual Heat Removal System (RHRS) of the Civaux NPP unit 1. An extensive investigation was carried out, and the conclusion was that the origin of this degradation phenomenon was cracking by thermal fatigue. The incident was caused by fluctuations in the temperature of the fluid downstream mixing tee. It is worth mentioning that the time between initiation of the crack and its development to a significant depth through the wall was only about $\approx$1500 hours, which is surprisingly low. Metallurgical examinations revealed substantial cracks and also some networks of small thermal fatigue cracks near the welds, but no fabrication defects. The section of interest is shown in Figure B1. The system operated at a pressure of 36 bars; the hot leg contains water at 180° C and the cold leg contains water at 20°C. In the damage zone of interest the pipe inner radius was $r_i \cong 120$ mm and outer radius were $r_o =129$ mm. The material properties are shown in Table B1.

Table B1 Thermal and mechanical properties of austenitic steel (304L) at room temperature

$c \left[\dfrac{J}{kg \cdot C}\right]$	$\lambda \left[\dfrac{W}{m \cdot C}\right]$	$\rho \left[\dfrac{kg}{m^3}\right]$	$\alpha \left[\dfrac{1}{C}\right]$	$E \left[\dfrac{N}{m^2}\right]$	v	$k \left[\dfrac{m^2}{s}\right]$
480	14.7	7800	$16.4.10^{-6}$	177.10^{9}	0.3	3.93×10^{-6}

c - specific heat coefficient, λ - thermal conductivity, ρ - density, α - mean thermal expansion,
k – thermal diffusivity coefficient.

The temperature fluctuation was reported to be in the range 20-180° C and on the inner surface of the pipe, the maximum temperature fluctuation range was estimated to be 120°C [3].

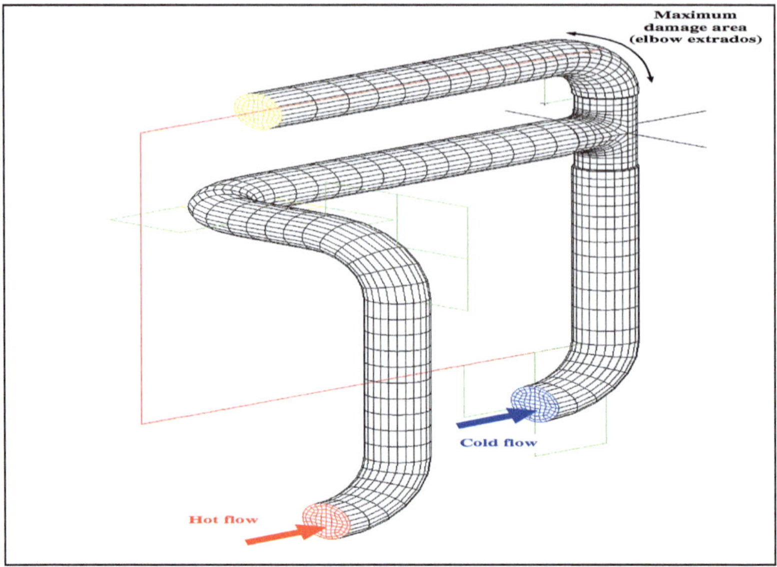

Fig. B1 The simplified sketch of piping subsystems with damaged area by thermal fatigue cracking [1]

References

[1] Chapuliot, S., Gourdin, C., Payen, T., Magnaud, J.P., Monavon, A.: Hydro-thermal-mechanical analysis of thermal fatigue in a mixing tee. Nuclear Engineering and Design 235, 575–596 (2005)

[2] : Nilsson, Assessment of thermal fatigue crack growth in the high cycle domain under sinusoidal thermal loading – An application-Civaux 1 case, published as European Commission Report EUR 23223 EN, DG JRC, Petten, The Netherlands (2007), doi:10.2790/4943 ISSN: 1018-5593, ISBN: 978- 92-79-08218-4

[3] Paffumi, E., Radu, V.: Status on the knowledge on cracks evolution under loadings from a thermal spectrum-Crack propagation and possible arrest/penetration, published as European Commission Report, European Communities (2009)

Appendix C
The First Hundred Roots of the Transcendental Equation (Civaux Pipe Geometry)

For Civaux 1 case, considering a pipe with the inner and outer radii $r_i \cong 0.120$ m (=0.1197m) and r_o=0.129 m, the first 100 roots of the transcendental Equation (A-13) used in the analytical solutions for temperature and stress fields are:

s_n =[337.78; 675.5900; 1.0134e+003; 1.3512e+003; 1.6890e+003;

2.0268e+003; 2.3646e+003; 2.7024e+003; 3.0403e+003; 3.3781e+003;

3.7159e+003; 4.0537e+003; 4.3915e+003; 4.7293e+003; 5.0671e+003;

5.4049e+003; 5.7427e+003; 6.0806e+003; 6.4184e+003; 6.7562e+003;

7.0940e+003; 7.4318e+003; 7.7696e+003; 8.1074e+003; 8.4452e+003;

8.7830e+003; 9.1208e+003; 9.4587e+003; 9.7965e+003; 1.0134e+004;

1.0472e+004; 1.0810e+004; 1.1148e+004; 1.1486e+004; 1.1823e+004;

1.2161e+004; 1.2499e+004; 1.2837e+004; 1.3175e+004; 1.3512e+004;

1.3850e+004; 1.4188e+004; 1.4526e+004; 1.4864e+004; 1.5201e+004;

1.5539e+004; 1.5877e+004; 1.6215e+004; 1.6553e+004; 1.6890e+004;

1.7228e+004; 1.7566e+004; 1.7904e+004; 1.8242e+004; 1.8580e+004;

1.8917e+004; 1.9255e+004; 1.9593e+004; 1.9931e+004; 2.0269e+004;

2.0606e+004; 2.0944e+004; 2.1282e+004; 2.1620e+004; 2.1958e+004;

2.2295e+004; 2.2633e+004; 2.2971e+004; 2.3309e+004; 2.3647e+004;

2.3984e+004; 2.4322e+004; 2.4660e+004; 2.4998e+004; 2.5336e+004;

2.5674e+004; 2.6011e+004; 2.6349e+004; 2.6687e+004; 2.7025e+004;

2.7363e+004; 2.7700e+004; 2.8038e+004; 2.8376e+004; 2.8714e+004;

2.9052e+004; 2.9389e+004; 2.9727e+004; 3.0065e+004; 3.0403e+004;

3.0741e+004; 3.1078e+004; 3.1416e+004; 3.1754e+004; 3.2092e+004;

3.2430e+004; 3.2768e+004; 3.3105e+004; 3.3443e+004; 3.3781e+004];

By using this set of values in the analytical solutions given by Equations (D1, D2, D3) from Appendix D, the corresponding sums will have the same number of terms.

Appendix D: Thermal Stress Components in an Infinite Hollow Cylinder under Sinusoidal Thermal Loading

The thermal stress components for a hollow circular cylinder subject to sinusoidal thermal loading are:

Hoop Thermal Stress

$$
\begin{aligned}
\sigma_\theta(r,\omega,t) = {}& \frac{\alpha \cdot E}{1-\nu} \times \left\{ \left(\frac{1}{r^2} \right) k \cdot \pi \cdot \sum_{n=1}^{\infty} \frac{s_n^2 \cdot J_0^2(s_n \cdot r_o)}{J_0^2(s_n \cdot r_o) - J_0^2(s_n \cdot r_i)} \times \right. \\
& \times \left[\frac{1}{s_n} \left\{ Y_o(s_n \cdot r_i) \cdot \left[r \cdot J_1(s_n \cdot r) - r_i \cdot J_1(s_n \cdot r_i) \right] - J_o(s_n \cdot r_i) \cdot \left[r \cdot Y_1(s_n \cdot r) - r_i \cdot Y_1(s_n \cdot r_i) \right] \right\} \right] \times \\
& \times \left[\theta_0 \cdot \frac{\omega \cdot e^{-k \cdot s_n^2 \cdot t} + (k \cdot s_n^2) \cdot \sin(\omega \cdot t) - \omega \cdot \cos(\omega \cdot t)}{(k \cdot s_n^2)^2 + \omega^2} \right] + \frac{r^2 + r_i^2}{r^2(r_o^2 - r_i^2)} \times \\
& \times k \cdot \pi \cdot \sum_{n=1}^{\infty} \frac{s_n^2 \cdot J_0^2(s_n \cdot r_o)}{J_0^2(s_n \cdot r_o) - J_0^2(s_n \cdot r_i)} \times \\
& \times \left[\frac{1}{s_n} \left\{ Y_o(s_n \cdot r_i) \cdot \left[r_o \cdot J_1(s_n \cdot r_o) - r_i \cdot J_1(s_n \cdot r_i) \right] - J_o(s_n \cdot r_i) \cdot \left[r_o \cdot Y_1(s_n \cdot r_o) - r_i \cdot Y_1(s_n \cdot r_i) \right] \right\} \right] \times \\
& \times \left[\theta_0 \cdot \frac{\omega \cdot e^{-k \cdot s_n^2 \cdot t} + (k \cdot s_n^2) \cdot \sin(\omega \cdot t) - \omega \cdot \cos(\omega \cdot t)}{(k \cdot s_n^2)^2 + \omega^2} \right] - \\
& - k \cdot \pi \cdot \sum_{n=1}^{\infty} \frac{s_n^2 \cdot J_0^2(s_n \cdot r_o)}{J_0^2(s_n \cdot r_o) - J_0^2(s_n \cdot r_i)} \left[Y_o(s_n \cdot r_i) \cdot J_o(s_n \cdot r) - J_o(s_n \cdot r_i) \cdot Y_o(s_n \cdot r) \right] \times \\
& \left. \times \left[\theta_0 \cdot \frac{\omega \cdot e^{-k \cdot s_n^2 \cdot t} + (k \cdot s_n^2) \cdot \sin(\omega \cdot t) - \omega \cdot \cos(\omega \cdot t)}{(k \cdot s_n^2)^2 + \omega^2} \right] \right\}
\end{aligned} \tag{D1}
$$

Axial Thermal Stress for $\varepsilon_z=0$ (Fixed End)

$$\sigma_z(r,\omega,t) = \frac{\alpha \cdot E}{1-v}\left\{ \frac{2v}{r_i^2 - r_o^2} \times k \cdot \pi \cdot \sum_{n=1}^{\infty} \frac{s_n^2 \cdot J_0^2(s_n \cdot r_o)}{J_0^2(s_n \cdot r_o) - J_0^2(s_n \cdot r_i)} \times \right.$$

$$\times \left[\frac{1}{s_n}\left\{ Y_o(s_n \cdot r_i) \cdot [r_o \cdot J_1(s_n \cdot r_o) - r_i \cdot J_1(s_n \cdot r_i)] - J_o(s_n \cdot r_i) \cdot [r_o \cdot Y_1(s_n \cdot r_o) - r_i \cdot Y_1(s_n \cdot r_i)] \right\} \right] \cdot \quad \text{(D2)}$$

$$\cdot \left[\theta_0 \cdot \frac{\omega \cdot e^{-k \cdot s_n^2 \cdot t} + (k \cdot s_n^2) \cdot \sin(\omega \cdot t) - \omega \cdot \cos(\omega \cdot t)}{(k \cdot s_n^2)^2 + \omega^2} \right] -$$

$$- k \cdot \pi \cdot \sum_{n=1}^{\infty} \frac{s_n^2 \cdot J_0^2(s_n \cdot r_o)}{J_0^2(s_n \cdot r_o) - J_0^2(s_n \cdot r_i)} \left[Y_o(s_n \cdot r_i) \cdot J_o(s_n \cdot r) - J_o(s_n \cdot r_i) \cdot Y_o(s_n \cdot r) \right] \times$$

$$\times \left[\theta_0 \cdot \frac{\omega \cdot e^{-k \cdot s_n^2 \cdot t} + (k \cdot s_n^2) \cdot \sin(\omega \cdot t) - \omega \cdot \cos(\omega \cdot t)}{(k \cdot s_n^2)^2 + \omega^2} \right] \Bigg\}$$

Axial Thermal Stress for $\varepsilon_z=\varepsilon_0$ (Free End)

$$\sigma_z(r,\omega,t) = \frac{\alpha \cdot E}{1-v}\left\{ \frac{2}{r_i^2 - r_o^2} \times k \cdot \pi \cdot \sum_{n=1}^{\infty} \frac{s_n^2 \cdot J_0^2(s_n \cdot r_o)}{J_0^2(s_n \cdot r_o) - J_0^2(s_n \cdot r_i)} \times \right.$$

$$\times \left[\frac{1}{s_n}\left\{ Y_o(s_n \cdot r_i) \cdot [r_o \cdot J_1(s_n \cdot r_o) - r_i \cdot J_1(s_n \cdot r_i)] - J_o(s_n \cdot r_i) \cdot [r_o \cdot Y_1(s_n \cdot r_o) - r_i \cdot Y_1(s_n \cdot r_i)] \right\} \right] \cdot \quad \text{(D3)}$$

$$\cdot \left[\theta_0 \cdot \frac{\omega \cdot e^{-k \cdot s_n^2 \cdot t} + (k \cdot s_n^2) \cdot \sin(\omega \cdot t) - \omega \cdot \cos(\omega \cdot t)}{(k \cdot s_n^2)^2 + \omega^2} \right] -$$

$$- k \cdot \pi \cdot \sum_{n=1}^{\infty} \frac{s_n^2 \cdot J_0^2(s_n \cdot r_o)}{J_0^2(s_n \cdot r_o) - J_0^2(s_n \cdot r_i)} \left[Y_o(s_n \cdot r_i) \cdot J_o(s_n \cdot r) - J_o(s_n \cdot r_i) \cdot Y_o(s_n \cdot r) \right] \times$$

$$\times \left[\theta_0 \cdot \frac{\omega \cdot e^{-k \cdot s_n^2 \cdot t} + (k \cdot s_n^2) \cdot \sin(\omega \cdot t) - \omega \cdot \cos(\omega \cdot t)}{(k \cdot s_n^2)^2 + \omega^2} \right] \Bigg\}$$

Appendix E: The Influence Coefficients G_j for Civaux Geometry and Dependence of Coefficients $H_j(\omega)$ on Loading Frequency

Table E1 displays the influence coefficients G_j ($j=0,1,2,3,4$) in the range of interest [1].

Table E1 The influence coefficients for a longitudinal infinite length surface crack in a cylindrical shell from Table C9 -API 579

r_i/l	a/l	G_0	G_1	G_2	G_3	G_4
5	0.0	1.12	0.682	0.5245	0.4404	0.379075
	0.2	1.307452	0.753466	0.564296	0.466913	0.398757
	0.4	1.8332	0.954938	0.676408	0.539874	0.454785
	0.6	2.734052	1.28757	0.857474	0.656596	0.54072
	0.8	3.940906	1.739955	1.10621	0.81823	0.661258
10	0.0	1.12	0.682	0.5245	0.4404	0.379075
	0.2	1.332691	0.763153	0.569758	0.470495	0.401459
	0.4	1.957764	1.002123	0.702473	0.556857	0.467621
	0.6	3.223438	1.466106	0.953655	0.718048	0.585672
	0.8	5.543784	2.300604	1.398958	1.000682	0.789201
20	0.0	1.12	0.682	0.5245	0.4404	0.379075
	0.2	1.345621	0.768292	0.57256	0.472331	0.402984
	0.4	2.028188	1.028989	0.717256	0.566433	0.475028
	0.6	3.573882	1.594673	1.023108	0.762465	0.618437
	0.8	7.388754	2.946567	1.736182	1.211533	0.936978

r_i – inner radius; l- wall-thickness.

The results of cubic spline interpolation (MATLAB function) for the influence coefficients in case of ratio $\dfrac{r_i}{l} \approx 13$ are shown in Table E2.

Table E2 Results of cubic spline interpolations for influence coefficients for the case of an infinite axial crack on inner pipe surface case

r_i/l	a/l	G_0	G_1	G_2	G_3	G_4
13	0.0	1.12	0.682	0.5245	0.4404	0.379075
13	0.2	1.3418	0.7667	0.5717	0.4718	0.4025
13	0.4	2.0039	1.0196	0.7121	0.5631	0.4724
13	0.6	3.4165	1.5367	0.9917	0.7424	0.6035
13	0.8	6.2878	2.5609	1.5349	1.0855	0.8487

Dependence of coefficients $h_j(\omega)$ on loading frequency.

$$h_0(f)=1.8950+11.5553\,f-26.4284\,f^2+25.9064\,f^3-9.3682\,f^4 \tag{E1}$$

$$h_1(f)=1.8950+11.5553\,f-26.4284\,f^2+25.9064\,f^3-9.3682\,f^4 \tag{E2}$$

$$h_2(f)=-45.1660+543.2349\,f-910.2986\,f^2+736.5647\,f^3-234.9905\,f^4 \tag{E3}$$

$$h_3(f)=10^3\,(\,0.0852-0.6924\,f+1.0106\,f^2-0.7131\,f^3+0.2013\,f^4) \tag{E4}$$

$$h_4(f)=-43.4634+281.9147\,f-341.3181\,f^2+184.2000\,f^3-35.7704\,f^4 \tag{E5}$$

Reference

[1] API 579 Fitness-for-Service-API Recommended Practice 579, 1st edn. American Petroleum Institute (January 2000)

Appendix F: The CDFs and PDFs Used for Probabilistic Input of Stochastic Assessment by Monte Carlo Method

a) Normal Distribution (Gauss Distribution)

Probability density function (PDF):

$$f(x;\mu,\sigma)=\frac{1}{\sigma\sqrt{2\pi}}e^{-\frac{1}{2}\left(\frac{x-\mu}{\sigma}\right)^2}\quad,\ -\infty<x<\infty;\ \ -\infty<\mu<\infty;\sigma2>0. \tag{F1}$$

Cumulative distribution function (CDF):

$$F(x;\mu,\sigma)=\frac{1}{\sigma\sqrt{2\pi}}\int_{-\infty}^{x}\exp\left[-\frac{(t-\mu)^2}{2\sigma^2}\right]dt \tag{F2}$$

Note: For $\mu=0$ and $\sigma=1$ we refer to this distribution as standard normal distribution

$$f(x)=\frac{1}{\sqrt{2\pi}}e^{-\frac{1}{2}x^2} \tag{F3}$$

which has the cumulative distribution (CFD)

$$\Phi(u)=\frac{1}{\sqrt{2\pi}}\int_{-\infty}^{u}\exp\left[-\frac{t^2}{2}\right]dt \tag{F4}$$

Moments:

mean (expected value):

$$E(X)=\mu_X=\mu. \tag{F5}$$

variance:

$$Var(X) = \sigma^2 \tag{F6}$$

standard deviation:

$$\sigma = \sqrt{Var(X)} \tag{F7}$$

b) Exponential Distribution

Probability density function (PDF):

$$f(x;\mu) = \frac{1}{\mu} e^{-\frac{x}{\mu}} \qquad x \geq 0;\ \mu > 0 \tag{F8}$$

Cumulative distribution function (CDF):

$$F(x;\mu) = 1 - e^{-\frac{x}{\mu}} \tag{F9}$$

Moments:
mean (expected value):

$$E(X) = \mu_X = \mu \tag{F10}$$

variance:

$$Var(X) = \mu^2 \tag{F11}$$

standard deviation:

$$\sigma = \sqrt{Var(X)} = \mu \tag{F12}$$

The exponential (Marshal) distribution is used to produce random value for initial crack depth a_i.

c) Log-Normal Distribution

Probability density function (PDF):

$$f(x;\mu_0,\sigma_0) = \frac{1}{x \cdot \sigma_0 \sqrt{2\pi}} e^{-\frac{1}{2}\left(\frac{\ln x - \mu_0}{\sigma_0}\right)^2} \tag{F13}$$

Cumulative Distribution Function (CDF):

$$F(x;\mu_0,\sigma_0) = \frac{1}{\sigma_0 \sqrt{2\pi}} \int_0^x \frac{1}{t} \cdot \exp\left[-\frac{(\ln(t) - \mu_0)^2}{2\sigma_0^2}\right] dt \tag{F14}$$

with $\mu_0,\ \sigma_0$, parameters.

Moments:

mean (expected value):

$$E(X) = \mu_X = e^{\mu_0 + \frac{\sigma_0^2}{2}} \tag{F15}$$

variance:

$$Var(X) = e^{2\mu_0 + \sigma_0^2}\left(e^{\sigma_0^2} - 1\right) \tag{F16}$$

standard deviation:

$$\sigma = \sqrt{Var(X)} \tag{F17}$$

median:

$$med = e^{\mu_0} \tag{F18}$$

The log-normal distribution is used to produce random values for C constant in Paris's law.